REVIEW OF THE

Research Program
of the

PARTNERSHIP

for a

NEW GENERATION

of

VEHICLES

FOURTH REPORT

Standing Committee to Review the Research Program
of the Partnership for a New Generation of Vehicles

Board on Energy and Environmental Systems
Commission on Engineering and Technical Systems
Transportation Research Board
National Research Council

NATIONAL ACADEMY PRESS
Washington, D.C. 1998

NATIONAL ACADEMY PRESS • 2101 Constitution Avenue, N.W. • Washington, D.C. 20418

This report and the study on which it is based were supported by Contract No. DTNH22-94-G-07414 from the National Highway Traffic Safety Administration, U.S. Department of Transportation. Any opinions, findings, conclusions, or recommendations expressed in this publication are those of the author(s) and do not necessarily reflect the view of the organizations or agencies that provided support for the project.

Library of Congress Catalog Number: 98-84940
International Standard Book Number: 0-309-06087-7

Available in limited supply from:

Board on Energy and Environmental Systems
National Research Council
2101 Constitution Avenue, N.W.
HA-270
Washington, DC 20418
202-334-3344

Additional copies are available for sale from:

National Academy Press
2101 Constitution Avenue, N.W.
Box 285
Washington, DC 20055
800-624-6242 or 202-334-3313 (in the Washington metropolitan area)
http://www.nap.edu

Acknowledgments

The committee wishes to thank all the members of the Partnership for a New Generation of Vehicles who contributed significantly of their time and effort to this National Research Council study, whether by giving presentations at meetings, providing responses to committee requests for information, or hosting site visits. The committee also acknowledges the valuable contributions of organizations outside the Partnership for a New Generation of Vehicles, including organizations outside the United States that provided information on advanced vehicle technologies and development initiatives. Finally, the chairman wishes to recognize the committee members and the staff of the Board on Energy and Environmental Systems of the National Research Council for their hard work organizing and planning committee meetings and for their individual efforts in gathering information and writing sections of the report.

This report has been reviewed by individuals chosen for their diverse perspectives and technical expertise, in accordance with procedures approved by the National Research Council's (NRC's) Report Review Committee. The purpose of this independent review is to provide candid and critical comments that will assist the authors and the NRC in making the published report as sound as possible and to ensure that the report meets institutional standards for objectivity, evidence, and responsiveness to the study charge. The content of the review comments and draft manuscript remain confidential to protect the integrity of the deliberative process. We wish to thank the follow individuals for their participation in the review of this report: William Agnew, General Motors Research Laboratories (retired); Tom Cackett, California Air Resources Board; Robert Frosch, Harvard University; John Heywood, Massachusetts Institute of Technology; John Longwell, Massachusetts Institute of Technology; Phillip Myers, University of

Wisconsin; Lawrence T. Papay, Bechtel Technology and Consulting; and Dale Stein, Michigan Technological University (retired).

    While the individuals listed above have provided constructive comments and suggestions, responsibility for the final content of this report rests solely with the authoring committee and the NRC.

Contents

Tables and Figures

TABLES

FIGURES

Executive Summary

This is the fourth report by the National Research Council's Standing Committee to Review the Research Program of the Partnership for a New Generation of Vehicles (PNGV). The PNGV program is a cooperative research and development program between the federal government and the United States Council for Automotive Research (USCAR), whose members are Chrysler Corporation, Ford Motor Company, and General Motors Corporation. One of the aims of the program, referred to as the Goal 3 objective, is to develop technologies for a new generation of vehicles that could achieve fuel economies up to three times (up to 80 mpg) those of comparable 1994 family sedans. At the same time, these vehicles should maintain performance, size, utility, and cost of ownership and operation and should meet or exceed federal safety and emissions requirements. The intent of the program is to develop concept vehicles by 2000 and production prototype vehicles by 2004. To meet this demanding schedule, a major PNGV milestone was the selection of the most promising technologies by the end of 1997 (sometimes referred to as the technology downselect).[1]

The committee's major tasks were to examine the overall balance and adequacy of the PNGV research program to meet the program goals and requirements (i.e., technical objectives, schedules, and rate of progress), examine the PNGV 1997 technology selection process, comment on the role of the government after the technology selection process and on how the PNGV program should interface, if appropriate, with other federal research programs. In assessing

---

[1]Goal 1 was to improve national competitiveness in manufacturing significantly, and Goal 2 was to implement commercially viable innovations from ongoing research on conventional vehicles. At the request of the PNGV, the sponsor of this study (see Appendix C), this report is focused on Goal 3 and the 1997 technology selection process.

progress and the efficacy of the program to meet its schedules and goals, the committee addresses several broad program issues. The committee also continues to review the PNGV systems analysis, which is essential (1) for making vehicle performance and cost comparisons for alternative vehicle configurations that incorporate widely different technology subsystem and component combinations and (2) for guiding the orderly selection and development of subsystem technologies with specific performance requirements for meeting the Goal 3 vehicle objectives.

This Executive Summary highlights the committee's principal findings and recommendations. More detailed recommendations in each of the areas addressed by the committee can be found in the body of the report. The major areas addressed in this summary are: (1) progress in research in each technical area, (2) the technology selection process, (3) the economic viability of the hybrid electric vehicle (HEV), (4) emission controls for the compression ignition direct injection (CIDI) engine, (5) the initiation of a comprehensive fuels strategy, (6) systems analysis, (7) safety, (8) the cost challenge, (9) adequacy and balance of the PNGV program, and (10) government involvement in post-concept vehicles.

PROGRESS IN RESEARCH

The PNGV Technical Roadmap details performance objectives and lays out milestones and schedules in each major technology area; the Roadmap has been updated for most of the PNGV technologies and provides a good summary of program goals. Although some technologies have now been dropped, the principal technology areas under development in the PNGV program are energy converters (CIDI engines and fuel cells), energy storage devices (batteries and flywheels), power electronics and electrical systems, and materials. In the committee's opinion, in spite of a shortfall in resources in many areas, good progress has been made in assessing the potential of each candidate system and identifying critical technologies necessary to make each system viable. In spite of this progress, however, the committee is concerned that the pace and funding of PNGV developments may not be at a level for the United States to remain competitive on an international basis. In early January 1998, individual news releases from USCAR partners recognized this possibility by announcing aggressive new technology programs that involve substantial investments. A combination of PNGV and in-house company and foreign developments no doubt provided an effective stimulus for the USCAR partners to move more aggressively.

Important technical advances during the past year are listed below:

- Internal Combustion Engines. Excellent progress has been made in the past year in all aspects of the four-stroke direct injection (4SDI) engine program, which has focused on the CIDI engine. A lightweight CIDI engine architecture study and a dimethyl ether (DME) alternative fuel design study were completed for CIDI engines. However, as is noted below, a recent stretch research objective for particulate emissions would require major changes in CIDI technology and fuels.

- Continuous Combustion Engines. As anticipated, the gas turbine and Stirling cycle engines have not reached a state of development suitable for use in the year 2000 concept vehicles, and both engines now fall into the post-PNGV technology development time frame. Progress has been made on the Stirling engine, except in the long-term retention and containment of the hydrogen working fluid; for gas turbines, the development of ceramic components has not progressed to the point that a low-risk, ceramic automotive gas turbine development program can be initiated. Nevertheless, the recent stretch research objective for particulate emissions may lead to the reevaluation of continuous combustion engines as potential energy converters.

- Fuel Cells. Significant accomplishments and excellent progress have been made on fuel cells. A carbon monoxide (CO) tolerant fuel cell stack (at a CO concentration of 50 ppm) and a partial oxidation gasoline processor integrated with a fuel cell stack for gasoline-to-electricity conversion were demonstrated. Cost is still a major issue, along with many other engineering developments.

- Electrochemical Energy Storage. Considerable progress has been made in the development of full-size cells of lithium-ion batteries and of nickel metal hydride batteries, although analysis via modeling is still at an early stage of development. One form of abuse tolerance was demonstrated for lithium-ion 6-Ah cells. Battery costs are still significantly high, and meeting the PNGV cost goals within the PNGV time frame will be an enormous challenge.

- Flywheels. Issues of safety, cost, and size are still barriers but are yielding to development programs. Recent data and design approaches have shown promise in overcoming a major safety issue, namely, containment of a flywheel failure for low-energy flywheels.

- Power Electronics and Electrical Systems. The PNGV technical team focused on this area has made considerable progress in organizing and coordinating its efforts, and a full-time leader has been appointed, as recommended by the committee in the Phase 3 report. Most targets for 1997 have been met or exceeded, and projected costs have been lowered substantially. An industry-wide specification was developed for an integrated power module. Meeting the PNGV cost targets is still a major challenge.

- Materials. The PNGV materials technology team and the vehicle engineering team have made a thorough evaluation of the lightweight candidate materials in preparation for the Goal 3 technology selection process. Weight-reduction potential has been determined for selected body structures. Aluminum is a leading candidate for the vehicle structure, and promising cost reduction initiatives are under way to bring the costs of an aluminum-intensive vehicle to a level competitive with today's steel vehicles. Glass-reinforced thermoplastic polymer is also a promising material for weight reduction because of the relatively low cost of the fiber, its fast cycle time, and its ability to integrate parts. Graphite-reinforced polymer is another candidate material but will require breakthroughs in cost and manufacturing techniques to be cost effective.

In addition to the advances listed above, systems analysis has been completed on fuel economy trade-offs for major vehicle configurations, and driveable hybrid propulsion test vehicles ("mules") are running at all three USCAR partners' development facilities.

TECHNOLOGY SELECTION PROCESS

During the first four years of the PNGV program, both the USCAR partners and the government research managers have examined hundreds of possible technologies and avenues that might contribute to the success of the program. The first major milestone for the program, targeted for the end of calendar year 1997, is the selection of the most promising technologies for the Goal 3 concept vehicles. The PNGV has reached its milestone for the initial technology selection process on schedule, and the USCAR partners can now continue with the design and construction of their year 2000 concept vehicles (see Appendix F). The committee notes and commends this progress, which is an important step toward meeting the demanding PNGV goals and schedule.

Each USCAR partner—Chrysler, Ford and General Motors—will develop separate concept vehicles, drawing from the spectrum of technologies developed under the PNGV. Based on the PNGV time frame and the level of maturity of the various technologies under consideration, the committee is not surprised that for the year 2000 concept vehicle the PNGV has focused on major weight reductions, reduced aerodynamic drag, reduced accessory loads, low-loss tires, CIDI engines, and a parallel HEV configuration that allows some level of recovery of braking energy. At the same time, the government will continue to work on high-risk enabling technologies that could be incorporated into subsequent concept vehicles between the year 2000 and 2004.

From the inception of the program and with limited resources, the PNGV program has not been able to bring alternative energy converters and storage devices—notably fuel cells, gas turbines, Stirling cycle engines, flywheels, and ultracapacitors—to the state of development at which they could be selected for the year 2000 concept vehicles. Furthermore, the committee found no evidence that the PNGV program has stimulated an increase in resources for the development of these alternative systems and devices for automotive applications, with the exception of a recent acceleration of U.S. investment in fuel cells. Observers who expected to see accelerated development of these higher-risk concepts by PNGV may be disappointed. The committee recognizes, however, that added resources might not have lowered the risk or increased the payoff of any of these emerging technologies to the point that they would have been selected. The committee also recognizes that PNGV has no agreed-upon schedule or specified levels of resources for the development of alternative technologies that were not selected for the initial (year 2000) concept vehicles. The committee anticipates that longer-term, possibly revolutionary advanced automotive technologies may emerge beyond the PNGV time frame.

Some of the year 2000 vehicle attributes will probably fall short of the established targets, but the preproduction prototypes are expected to meet the targets by 2004. Given the potential uncertainties in performance, utility, and cost of the concept vehicle, PNGV established the following criteria for its technology selection:

- Fuel Economy. The vehicle must have a fuel economy of up to 80 mpg.
- Emissions. The vehicle must meet the Tier II Clean Air Act Amendments gaseous emission standards and applicable ultra-low emission particulate standards.
- Safety. The vehicle design must meet federal motor vehicle safety standards.
- Performance. The vehicle should be within ± 30 percent of Goal 3 targets.
- Utility.[2] The vehicle should be within ± 30 percent of Goal 3 targets.
- Cost Potential. The cost should be within 30 percent of the cost of the baseline vehicle.

Table ES-1 lists the most promising technologies chosen by the PNGV in 1997. PNGV will discontinue development of gas turbine engines, Stirling engines, and ultracapacitors as energy storage devices. Based on calculations and assessments, PNGV computed the relative potential fuel economy of a number of vehicle power train configurations (see Figure ES-1). The projected fuel economies of these configurations range from 27 mpg to more than 80 mpg. The priority will be to reduce emissions from CIDI engines through combustion development, after-treatment, and fuel modifications. The development of a low-cost, low-mass vehicle, a low-cost, high-efficiency electric drive, and fuel cells will also be priorities.

Excellent progress has been made in the development of CIDI engines. PNGV has identified a stretch research objective[3] of 0.01 g/mile for particulate emissions that poses a significant challenge, in addition to the already difficult

---

[2]Utility refers to the degree to which a given vehicle is useful to the individual car buyer and includes such attributes as passenger space, trunk capacity, seating capacity, and ergonomics.

[3]A level of 0.01 g/mile is approximately equivalent to gasoline engine emissions.

TABLE ES-1   Most Promising Technologies Selected by PNGV in 1997

Category	Technical Area and/or Technology
Power trains	parallel hybrid electric drive
Energy converters	CIDI engine fuel cells
Energy storage	nickel metal hydride batteries lithium batteries
Emission controls	lean NO_x catalyst exhaust gas recirculation particulate traps
Fuels	fuel with less than 50 ppm sulfur Fischer-Tropsch fuel dimethyl ether fuel
Electrical systems and electronics	induction, reluctance, permanent magnet motors PEBB, IGBT, MOSFET, MCT semiconductors ultracapacitors
Materials	aluminum and/or reinforced polymer body-in-white
Reduced energy losses	low rolling resistance tires reduced HVAC requirements and more efficient components

Source: Based on York (1997) and the PNGV Technology Selection Announcement (see Appendices D and F).

Acronyms: PEBB = power electronic building block; HVAC = heating, ventilation, and air conditioning; IGBT = insulated gate bipolar transistor; MOSFET = metal oxide semiconductor field effect transistor; MCT = MOS (metal oxide semiconductor) controlled thyristor. Body-in-white constitutes the primary structural frame of the vehicle not including bolt on pieces, such as the hood, doors, front fenders, and deck lids.

problem of reducing nitrogen oxide (NO_x) emissions. The prospects for developing a CIDI engine that would meet this stretch objective would change from encouraging to high risk. Given the stretch emissions objective, PNGV will need to reevaluate other candidate engines and system configurations. For example, Mitsubishi has published and reported to the committee that the fuel economy for its gasoline direct-injection engine may approach the fuel economy of the CIDI engine while achieving low-emission vehicle standards. As the PNGV program moves into the concept vehicle development phase, priorities among different technologies must be established.

Recommendation. In light of the published improvements in gasoline direct-injection engines, it would be prudent for the PNGV partners to continue to assess developments in this technology against PNGV targets and the CIDI

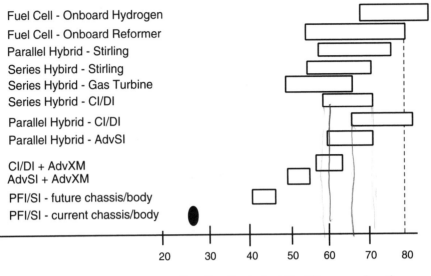

Gasoline Equivalent Fuel Economy (mpg)

FIGURE ES-1   Relative fuel economy projections for various vehicle/power train configurations. All configurations except the current chassis/body include PNGV-class light-weight body, chassis, and interior (2,000 lbs); advanced aerodynamics; and low rolling resistance tires. The PFI/SI (port fuel injection/spark ignition) conventional vehicle is the baseline. Key: ● = current vehicle chassis/body; □ = future efficient vehicle body/chassis [light weight (2,000 lbs), sleek aerodynamics, and low rolling resistance]. Variance denotes downward uncertainty from unmodeled energy losses and upward uncertainty from improvement as technology matures. CIDI = compression ignition/direct injection; AdvSI = advanced spark ignition. AdvXM = advanced transmission. Source: Provided to the committee by PNGV.

engine, whether or not the gasoline direct-injection engine is chosen as a potential PNGV power plant.

Recommendation. The relationship between the criteria for technology selection and the critical requirements of Goal 3 should be made more explicit to facilitate the proper distribution of resources for an ongoing, well structured research and development program.

ECONOMIC VIABILITY OF THE HYBRID ELECTRIC VEHICLE

Although significant progress continues to be made in technology development, the economic viability of the HEV remains to be demonstrated. HEVs are more complex than nonhybrid vehicles because of their added energy storage and recovery devices, electrical drive, and electronic controls for electric power. Despite the projected improvement in fuel economy of an HEV with a CIDI engine over a nonhybrid vehicle, the committee believes the difference is probably not enough to offset the higher cost of the HEV.

Recommendation. The PNGV should continue to refine its detailed cost of ownership analyses of hybrid vs. nonhybrid vehicles. If the economic and performance benefits of the hybrid vehicle do not exceed or warrant its additional costs, the concept demonstration vehicle program should be expanded to include nonhybrid vehicles to accelerate the development and commercial introduction of economically viable technologies.

EMISSIONS CONTROL FOR COMPRESSION IGNITION DIRECT INJECTION ENGINES

The committee (and the PNGV 4SDI engine team) noted in its Phase 3 report that the most technically challenging aspect of the CIDI engine program will be meeting the NO_x and particulate emission standards. In addition, meeting a more severe particulate matter emissions stretch research objective (0.01 g/mile) instead of the 0.04 g/mile target would require additional technological breakthroughs for the CIDI engine to meet PNGV milestones.

Achieving these low emission levels will require significant advances in combustion control, as well as exhaust after-treatment and fuel composition changes. The extent to which combustion can be controlled through the use of electronic, high-pressure, fuel injection systems in small diesel engines is not known, but ongoing cooperation between the PNGV and the heavy-duty diesel engine industry should help to address this issue.

Another very challenging area is exhaust-gas after-treatment, which will be required both for NO_x and particulate emissions for CIDI engines to meet the PNGV goals. Both lean NO_x catalysis and plasma-assisted after-treatment are being investigated, but progress in both areas has been slow. Despite the significant technical challenges, PNGV plans to continue to focus on exhaust-gas after-treatment.

The PNGV recognizes the importance of the interaction between fuel properties and engine performance and emissions and is involved in programs to assess the potential for reducing emissions through fuel changes. For CIDI engines to meet the existing target of 0.2 g/mile for NO_x and a research objective of 0.01 g/mile for particulate matter, PNGV will have to focus on alternative fuels.

Recommendation. The PNGV should devote considerably more effort and resources to the exhaust-gas after-treatment of oxides of nitrogen and particulates. PNGV should consider greatly expanding its efforts to involve catalyst manufacturers.

INITIATION OF A COMPREHENSIVE FUELS STRATEGY

As the committee noted in previous reports, the PNGV Goal 3 is focused almost exclusively on vehicle issues, and no consensus has been reached on how to weigh trade-offs between the energy consumed or emissions produced by the vehicle and the energy consumed or emissions produced throughout the fuel processing and distribution infrastructure. Fuels will be critical to PNGV meeting the emission targets for the CIDI engine; for example, lowering the sulfur content of the fuel will lead to lower particulate emissions. Sulfur levels will also have to be low for a fuel cell with an onboard petroleum reformer. Modifying current refinery processes or synthesizing liquid fuels from natural gas could produce low-sulfur fuels. Widespread distribution of hydrogen will be a major infrastructure issue in the practicality of a fuel cell powered by hydrogen carried on the vehicle.

An ample supply of properly tailored fuel will be critical to the success of any automotive energy converter. If a fuel significantly different from the ones now in use will be needed for future automobiles, extensive consideration of the feasibility and economics associated with its production and distribution should be undertaken early in the development process. If a major change in the fuel system infrastructure is required, even longer lead-time will be required. If the PNGV determines that a sequence of changes will be necessary, the economics of the industries involved should be studied carefully. The study may indicate that only one major change is economically feasible for the foreseeable future. If so, PNGV's choice of automotive energy converters will be restricted to converters that are compatible with a fuel that could be used for all of them.

Recommendation. The PNGV should propose ways to involve the transportation fuels industry in a partnership with the government to help achieve PNGV goals.

Recommendation. PNGV's choices of energy conversion technologies should take full account of the implications for fuel development, supply, and distribution (infrastructure), as well as the economics and timing required to ensure the widespread availability of the fuel.

SAFETY

New structural materials, power plants, fuels (including hydrogen), energy storage devices, and glazing materials are being considered by the PNGV to improve power train efficiency, energy storage, and weight reduction. Each new technology is likely to introduce new failure modes and new safety concerns in crash performance, flammability, potential for explosions, electrical shock, and toxicity. The committee decided not to review safety issues in depth with the PNGV technical teams until after the technology selection milestone had been reached; however, the committee is satisfied that the technical teams are aware of these issues and are addressing them on an ongoing basis as part of the overall program. For example, failure modes are under investigation for all promising technologies; the issues associated with handling and storing onboard hydrogen for fuel-cell-powered vehicles are being addressed; and computer simulations are being used to examine the crash performance of hybrid vehicles. In light of the role of the U.S. Department of Transportation-National Highway Traffic Safety Administration to regulate vehicle safety and its recent study on vehicle size and weight as related to safety, the committee believes it is appropriate that the National Highway Transportation Safety Administration become involved in studies of the crashworthiness of lightweight vehicles comparable to PNGV designs.

Recommendation. PNGV and the USCAR partners should continue to make safety a high priority as they move toward the realization of the concept vehicles.

SYSTEMS ANALYSIS

The committee has been very critical in past reports of the lack of progress and timeliness of the PNGV systems analysis, which is critical to analyzing vehicle engineering trade-offs and to setting the performance and cost priorities of subsystem technologies. The committee is pleased to see that considerable progress has been made in the past year and that efforts are under way to provide effective design support to the PNGV technical teams through systems analysis.

As the technology selection process for the concept vehicles continues between now and the year 2000, as well as beyond 2000, and as second-generation concept vehicles emerge and production prototypes are developed, the need for a robust and strong systems engineering and analysis team will increase. At this point, the PNGV has done little to incorporate cost modeling, and the committee believes that probabilistic models of vehicle and subsystem costs, with confidence levels, will be important tools for making decisions in the future. Another area that deserves more attention is reliability studies because every new subsystem being considered for the Goal 3 vehicles introduces new failure modes, which should be evaluated.

Recommendation. Systems analysis and computer modeling are essential tools for making system trade-offs and optimizing performance. The PNGV should create detailed, rigorous cost and design reliability models as soon as possible to support ongoing technology selection. These models should be continuously upgraded as new information becomes available.

FUEL CELL DEVELOPMENT

Of the known technologies, fuel cells have the best long-term potential for automotive energy converters with high efficiency and low emissions; several impressive advances have been made in the performance of fuel cell stacks and fuel-to-hydrogen reformers. Nevertheless, fuel cells still face substantial obstacles to meeting performance and cost goals within the 2000 to 2004 time frame.

Interesting developments have occurred both abroad and in the private sector. The most visible foreign developments have been made by Ballard Power Systems, Inc., of Canada, and Daimler-Benz, of Germany. A $450 million (Canadian) joint venture between these two companies has been formed to develop fuel-cell technology for vehicles. Ford Motor Company announced in the fall of 1997 that it would join this partnership with an investment of $420 million in cash, technology, and assets. The committee believes significant government-industry investments in fuel-cell technologies for transportation are also being made elsewhere in Europe and in Japan. Given this worldwide interest, the committee believes the PNGV should continue to make substantial efforts and investments in this potential automotive energy converter.

Recommendation. Because of their high efficiency and low emission potential, fuel-cell systems for transportation could become vitally important to the United States. U.S. government and industry investments in research and development should, therefore, be continued at current levels or even be increased for an extended period.

THE COST CHALLENGE

Although significant technical progress has been made in many of the technologies that will be incorporated into the PNGV Goal 3 vehicles, meeting the cost targets remains a formidable challenge. A review of the proprietary cost models and analyses of two of the USCAR partners confirmed the magnitude of the Goal 3 cost challenge. The committee members who reviewed the cost data were fully satisfied that substantial in-depth cost analyses were being performed by two of the USCAR partners and that these analyses had influenced their product designs. The results of these analyses should be communicated to the PNGV systems analysis team for use in their cost modeling efforts without compromising the USCAR partners' proprietary interests.

Recommendation. Because cost is a significant challenge to PNGV, the USCAR partners should continue to conduct in-depth cost analyses and to use the results to guide new development initiatives on components and subsystems.

ADEQUACY AND BALANCE OF THE PNGV PROGRAM

The committee found it difficult to assess the efforts and resources applied to the PNGV program because no funding plan was made available. However, the program to date has supported the selection of technologies for the concept vehicles with high potential for approaching the 80 mpg fuel economy target within the

PNGV time frame. Although it is estimated to fall short of the 80 mpg goal, the projected fuel economy will represent a major achievement for either hybrid or nonhybrid vehicle configurations. Advances in HEV components and the work done to meet goals 1 and 2, as well as industry's attempts to reduce vehicle weight by using aluminum and other lightweight materials, are improving the prospects that PNGV vehicles will have high fuel efficiency. Assuming that the PNGV/ USCAR partners perform as expected, that attempts to reduce costs are successful, and that the formidable challenge of emission requirements can be overcome, the committee believes that PNGV has allocated adequate resources to the selected technologies to realize, with a high degree of confidence, the year 2000 concept demonstration vehicles and the 2004 production prototype vehicles.

In the future, the committee expects to see substantial progress toward the production and commercial introduction of the lower-risk technologies embodied in the concept vehicles, as well as progress toward overcoming the barriers encountered in the post-concept vehicle technologies. Increased attention worldwide to carbon dioxide levels, and the potential that an international agreement will require implementation of a national strategy, could accelerate the necessity of introducing lower risk near-term improvements in fuel consumption into the world's automotive fleets. Nonhybrid vehicles being developed globally use relatively low-risk technologies and have a potential to reduce the cost of ownership. Without minimizing the risk and the substantial efforts, large capital investment, and high infrastructure costs that will be required for further development, the committee notes the potential for the relatively early implementation of these technologies. A strategy of accelerated implementation of the technologies leading to an approximately 60+ mpg nonhybrid vehicle, coupled with an expanded research plan to reach the 80 mpg target, seems both prudent and feasible.

Another committee concern is ensuring that technology developed for the PNGV midsize sedan is appropriately used for light trucks (pickups, minivans, and sport utility vehicles). The current light-truck market share has increased to almost 50 percent of sales. These vehicles are heavier and consume more fuel per mile than the average automobile. If present trends continue, making an impact on total U.S. transportation fuel consumption through the PNGV will require greater attention to light trucks.

Recommendation. Government and industry policy makers should review the benefits and implications of PNGV pursuing a parallel strategy to achieve a 60+ mpg nonhybrid vehicle at an early date and should establish goals, schedules, and resource requirements for a coordinated development program.

Recommendation. The PNGV should assess the implications of the growing vehicle population of light trucks in the U.S. market in terms of overall fuel economy, emissions, and safety. Wherever possible, the PNGV should develop strategies for transferring technical advances to light trucks.

GOVERNMENT INVOLVEMENT IN THE
DEVELOPMENT OF POST-CONCEPT VEHICLES

The committee was asked to comment on the role of government beyond the 1997 technology selection process and on how the PNGV might interact with other government research programs. Separate concept demonstration vehicles (by 2000) and production prototypes (by 2004) will be built by the USCAR partners with no significant participation by government. However, the government can and should support the development of longer-term technologies that are likely to be incorporated in subsequent concept vehicles. Beyond the demonstration of the concept vehicles and the PNGV time frame, the government should take the lead in developing high-risk, long-term technologies for vehicles with low fuel consumption and emissions. The development of these advanced vehicles will be especially important in light of concerns about climate change, concerns about maintaining U.S. competitiveness, and concerns about the country's balance of payments.

The PNGV is a partnership of seven government agencies and the three USCAR partners, but it does not have the line management structure or budgetary authority to control projects by different agencies that may have different missions. One of PNGV's primary functions is to establish communications and the exchange of information on the technology and projects related to automobiles and to coordinate recommendations and jointly plan future projects. The PNGV should see that the public gets the maximum benefit from government-funded development by the various agencies and should encourage the support of high payoff technologies in automotive applications, such as low emissions, high efficiency, and low cost of ownership.

Recommendation. The government should significantly expand its support for the development of long-term PNGV technologies that have the potential to improve fuel economy, lower emissions, and be commercially viable.

Recommendation. The PNGV should expand its liaison role for the exchange of technological information among federal research programs that are relevant to automotive technologies and should accelerate the sharing of results among the participants in the PNGV on long-term, high-payoff technologies applicable to automobiles.

1

Introduction

On September 29, 1993, President Clinton initiated the Partnership for a New Generation of Vehicles (PNGV) program, which is a cooperative research and development (R&D) program between the federal government and the United States Council for Automotive Research (USCAR), whose members are Chrysler Corporation, Ford Motor Company, and General Motors Corporation (GM).[1] The purpose of the PNGV program is to improve substantially the fuel efficiency of today's automobiles and enhance the U.S. domestic automobile industry's productivity and competitiveness. The objective of the PNGV program over the next decade is to develop technologies for a new generation of vehicles that can achieve fuel economies up to three times (80 miles per equivalent gallon of gasoline) those of today's comparable midsize sedans, while maintaining performance, size, utility, and cost of ownership and operation and meeting or exceeding federal safety and emissions requirements (The White House, 1993).

The PNGV declaration of intent includes a requirement for an independent peer review "to comment on the technologies selected for research and progress made." In response to a written request by the undersecretary for technology administration, U.S. Department of Commerce, acting on behalf of the PNGV, the National Research Council (NRC) in July 1994 established the Standing Committee to Review the Research Program of the Partnership for a New Generation of Vehicles. The committee conducts annual independent reviews of the PNGV's

---

[1]USCAR, which predated the formation of PNGV, was established to support intercompany precompetitive cooperation aimed at reducing the cost of redundant R&D in the face of intracompany international competition. USCAR is currently comprised of a number of consortia, as shown in Appendix E.

research program, advises the government and industry participants on the program's progress, and identifies significant barriers to success. This is the fourth report by the committee; the previous three are documented in three NRC reports, which provide further background on the PNGV program and committee efforts (NRC, 1994, 1996, 1997).

The PNGV goals and the considerations underlying all of the NRC reviews articulated in the partnership's program plan are listed below (PNGV, 1995):

Goal 1. Significantly improve national competitiveness in manufacturing for future generations of vehicles. Improve the productivity of the U.S. manufacturing base by significantly upgrading U.S. manufacturing technology, including the adoption of agile and flexible manufacturing and reduction of costs and lead times, while reducing the environmental impact and improving quality.

Goal 2. Implement commercially viable innovations from ongoing research on conventional vehicles. Pursue technology advances that can lead to improvements in fuel efficiency and reductions in the emissions of standard vehicle designs, while pursuing advances to maintain safety performance. Research will focus on technologies that reduce the demand for energy from the engine and drive train. Throughout the research program, the industry has pledged to apply those commercially viable technologies resulting from this research that would be expected to increase significantly vehicle fuel efficiency and improve emissions.

Goal 3. Develop vehicles to achieve up to three times the fuel efficiency of comparable 1994 family sedans. Increase vehicle fuel efficiency to up to three times that of the average 1994 Concorde/Taurus/Lumina automobiles with equivalent cost of ownership adjusted for economics.

As the committee noted in its third report, significant improvements in automotive fuel economy and the development of competitive advanced automotive technologies and vehicles can provide important economic benefits to the nation. The automotive industry, which is an important component of the U.S. economy, can benefit greatly if it can meet the expected increase in demand for cost-effective, fuel-efficient products in international markets, especially in Asia. Highway traffic accounts for a significant part of the atmospheric ozone in urban areas, contributes to greenhouse gas emissions, and is largely responsible for U.S. petroleum consumption and the consequent outflow of dollars for the purchase and import of petroleum products (NRC, 1997; OTA, 1995; Sissine, 1996).

Several studies completed in 1997, as well as President Clinton's Climate Change Proposal, have focused on significant improvements in energy efficiency in various sectors of the economy that could be realized through energy R&D on technologies for energy production and energy consumption. Important reductions in greenhouse gas emissions could also result from the development of advanced technologies (DOE, 1997; PCAST, 1997). The transportation sector of the economy figures prominently in these studies.

Currently, U.S. gasoline prices are relatively low, so automobile purchasers have little incentive to consider fuel economy in their purchase decisions. The U.S. automotive market has experienced a substantial increase in the sales of light trucks, especially of sport utility vehicles, which have substantially lower fuel economy requirements than automobiles. If present trends continue, light trucks will dominate the market sometime between 2004 and 2008 (DOE, 1996; NRC, 1992; NRC, 1998). The lack of market incentives for car buyers to purchase vehicles with high fuel economy makes the public benefits from improvements in fuel economy, such as the health benefits from reduced urban ozone, "insurance" against sudden increases in crude oil prices, the lower costs of maintaining energy security, the potential savings from lower crude oil prices, the improved balance of payments, and the reductions in greenhouse gases from the transportation sector, difficult to realize (Sissine, 1996; OTA, 1995).

The PNGV strategy of developing an automobile with a fuel economy of up to 80 mpg, and maintaining performance, size, utility, and cost and meeting or exceeding safety and emissions standards, circumvents the lack of economic incentives for buying automobiles with high fuel economy. If the PNGV strategy is successful, the buyer will purchase a vehicle with all of the desirable consumer attributes, as well as greatly enhanced fuel economy. The development of this vehicle, as the committee noted in its previous reports, is extremely challenging. But this ambitious goal will stimulate the rapid development of the required technologies and, even if a Goal 3 vehicle does not achieve the triple-level fuel economy while approaching cost and performance objectives, it may still be far more fuel-efficient than current vehicles, which would be an outstanding achievement.

In the past few years, but especially in the last year, a number of events related to advanced automotive technologies for passenger vehicles that are relevant to the PNGV program have occurred, including: (1) limited consumer interest in the EV1, the GM electric vehicle introduced into the market in 1996; (2) large investments announced by Daimler-Benz ($450 million Canadian) and Ford Motor Company ($420 million) in joint ventures with Ballard Power Systems, Inc., a Canadian fuel cell manufacturing firm, to develop fuel-cell technology; (3) the introduction in Japan of the Prius by Toyota, the first commercial hybrid vehicle (fuel economy of about 60 mpg); (4) recent regulations by the Environmental Protection Agency (EPA) to reduce atmospheric concentrations of particulate matter; (5) an agreement in February 1998 to establish a National Low Emission Vehicle program; (6) recent identification by PNGV of a stretch research objective of 0.01 g/mile for particulate matter emissions; and (7) the display at the 1997 Tokyo Motor Show by Japanese automotive companies of an array of production-ready "green" vehicles, i.e., vehicles with low emissions, high fuel economy, and high recyclability.

Competition fostered by the PNGV has no doubt hastened many of these overseas developments. For example, the European Car of Tomorrow Task Force was initiated in 1995, and the Japan Clean Air Program was launched in 1996. These events highlight the potential strategic value of programs like the PNGV and the importance of staying abreast of developments in foreign technology.

The PNGV's objective is to bring together the extensive R&D resources of the federal establishment, including the national laboratories and university-based research institutions, and the vehicle design, manufacturing, and marketing capabilities of both the USCAR partners and suppliers to the automotive industry. In general, government funding for the PNGV is primarily used for the development of longer-term, high-risk technologies. Funding by USCAR and industry is generally used to develop technologies with nearer-term commercial potential, to implement government technology developments in automotive applications, and to produce concept vehicles. Substantial in-house proprietary R&D programs are also ongoing at USCAR partners' facilities.

Technical teams responsible for R&D on the candidate subsystems, such as fuel cells, gas turbines, compression ignition direct injection (CIDI) engines, and others, are central to the PNGV. A manufacturing team, an electrical and electronics power conversion devices team, a materials and structures team, and a systems analysis team are also part of the PNGV organization (NRC, 1996, 1997). Technical oversight and coordination are provided by the vehicle engineering team, which provides the technical teams with vehicle system requirements, which are supported by the systems analysis team.

This review, the fourth by the committee, is being conducted during the critical process of technology selection (often referred to as the technology "downselect" process) by the end of 1997. According to the schedule for Goal 3 described in the PNGV Program Plan, the PNGV was to assess system configurations for alternative vehicles and to narrow its technology choices by the end of 1997, with the intent of defining, developing, and constructing concept vehicles by 2000 and production prototypes by 2004 (PNGV, 1995). The USCAR partners will develop separate concept vehicles, drawing on the spectrum of technologies developed under PNGV and in-house proprietary technology. PNGV itself will not design or build a concept car, a decision that the committee supports. Although the 1997 technology selection process focused on choosing the technologies most likely to result in concept and production prototype vehicles that could meet the Goal 3 requirements, other longer-range technologies will continue to evolve and may be incorporated into subsequent concept vehicles, as appropriate. As important technological advances are made, a series of concept vehicles will probably be developed beyond the year 2000. Since the beginning of the program, the PNGV has addressed many technology areas, including advanced lightweight materials and structures; efficient energy conversion systems (including advanced internal combustion engines, gas turbines, Stirling engines, and fuel cells); hybrid electric propulsion systems; energy-storage devices (including high-power batteries, flywheels, and ultracapacitors); efficient electrical and electronic systems; and systems that can efficiently recover and utilize exhaust energy and braking energy.

This fourth PNGV review was conducted by a committee comprising 15 members with a wide variety of expertise (see Appendix A for biographical information). Given the 1997 milestone of technology selection for the Goal 3 vehicle, the PNGV asked that the committee focus attention on progress made toward meeting Goal 3 and the development of relevant technologies. The objectives of goals 1 and 2, in many instances, support progress toward Goal 3, especially development of the manufacturing capabilities for the advanced automotive technologies being considered for the Goal 3 vehicle. The committee was asked to address the following tasks in this review:

- In light of major technical accomplishments since the third review and technical barriers that remain to be overcome, as well as the response by the PNGV to previous committee recommendations, examine the overall balance and adequacy of the PNGV research effort to meet the program goals and requirements, i.e., technical objectives, schedules, and rate of progress necessary to meet these requirements.
- Examine the PNGV technology selection process including how the PNGV is making choices and the role of government in the PNGV program after the technology selection process is completed.
- Consider and comment on how the PNGV program should interface, if appropriate, with other federal research programs.
- Prepare a fourth peer review report.

In view of the importance of Goal 3 at this point in the PNGV program, the committee was directed not to review goals 1 and 2 (see Appendix C) and not to review or analyze the resources being applied to the PNGV overall (see appendices B and C).

The fourth peer review report contains the committee's conclusions and recommendations (Appendix B contains a list of meetings, presentations, and other data-gathering activities by the committee).[2] Some of the material reviewed by the committee was presented by USCAR as proprietary information under an agreement signed by the National Academy of Sciences, the USCAR, and the U.S. government (represented by the U.S. Department of Commerce).

As the committee noted in previous reports (NRC, 1996, 1997), all of these reviews have been undertaken with the understanding that the vision, goals, and

---

[2]The committee formed the following subgroups: Nonelectrochemical Storage Devices; Electrical and Electronic Systems; Systems Analysis; Batteries and Ultracapacitors; Fuel Cells; Internal Combustion Engines; Continuous Combustion Engines; Materials; Fuels; Vehicle Technology Selection; and Cost Analysis. For a list of members of the subgroups see page iii.

target dates for the PNGV had been articulated by the President and that the appropriate R&D programs had been launched. On the assumption that the PNGV partners will seriously pursue the objectives of the program, the committee understands its charge as providing independent advice to help the PNGV achieve its goals. Therefore, the committee has tried to identify actions that could enhance the program's chances for success. The committee continues to avoid making judgments on the value of the program to the nation and accepts the goals as given, noting that goals 1 and 2, unlike Goal 3, are open-ended and do not have quantitative targets and milestones. The committee's objective continues to be to review the R&D program as presently configured and to assess the PNGV's progress toward, and potential for, achieving its goals. However, because regulatory and market changes occur continually, the PNGV should keep abreast of relevant changes and reassess its objectives.

2

Development of Vehicle Subsystems

CANDIDATE SYSTEMS

The ultimate success of the PNGV program will be measured by its ability to integrate R&D programs that collectively improve the fuel efficiency of automobiles within the very stringent boundary conditions of size, reliability, durability, safety, and affordability of today's cars. At the same time, the vehicles must meet even more stringent emission and recycling levels and must use components that can be mass produced and maintained in a manner similar to current automotive products.

In order to achieve a Goal 3 fuel economy that approaches the 80 mpg target (80 mpg is about three times the fuel efficiency of today's comparable vehicles), the energy conversion efficiency of the chemical conversion system (e.g., a power plant, such as a CIDI engine, a gas turbine, a Stirling engine, or a fuel cell) averaged over a driving cycle will have to be at least 40 percent, approximately double today's efficiency. This is an extremely challenging goal and will require assessing many possible concepts for improving efficiency. For example, the PNGV high fuel economy level of 80 mpg will require the integration of the primary power plant with energy storage devices, as well as the use of lightweight materials for the vehicle structure to reduce vehicle weight.

Despite concerted efforts in the last year to develop and evaluate the various candidate systems, none of the energy conversion power trains being considered meets all of the constraints. Therefore, R&D programs on both the selected and nonselected candidate systems have to be continued after the 1997 technology selection for the first concept vehicles, with the objective of attaining the breakthroughs that would make one or more of the technologies viable for meeting Goal 3 requirements. In 1997, the PNGV identified a stretch research objective of

0.01 g/mile for emissions of particulate matter for the CIDI engine. The current target is 0.04 g/mile. Meeting the stretch research objective presents new challenges to the candidate CIDI engine, which would require expanded technology development to meet the PNGV goals. PNGV would also need to reevaluate other power plants relative to CIDI engines.

The hybrid electric vehicle (HEV), which is the PNGV power train of choice, uses an energy storage device to decrease the fluctuations in the demands on the primary power plant. This reduction allows for a decrease in the peak power output required from the primary energy conversion system and an opportunity to improve efficiency both by restricting the power fluctuations and by recovering a significant fraction of the vehicle's kinetic energy during braking operations. The PNGV is sponsoring research on batteries, flywheels, and ultracapacitors as energy storage devices.

Achieving the high fuel economy levels for the Goal 3 vehicle will require more than improving the energy conversion efficiency of the power train (including energy converters and transmissions) and reducing other energy losses in the vehicle. Vehicle weight reduction through the use of new vehicle designs and lightweight materials will be extremely important in achieving the very ambitious fuel economy targets.

The committee re-evaluated the candidate energy conversion and energy storage technologies, as well as candidate electrical and electronic systems, that were considered last year and addresses them in this chapter. This chapter also reviews progress on advanced structural materials for the vehicle body, a subject that was not addressed by the committee in its third review. The technologies evaluated in this chapter are listed below:

- four-stroke CIDI engines
- continuous combustion systems
- fuel cells
- electrochemical storage systems
- electro-mechanical storage systems
- electrical and electronic power-conversion devices
- materials

The committee reviewed R&D programs on each of these technologies to assess the progress that has been made and the developments required for the future. The PNGV Technical Roadmap, which has been updated for most of these technologies, provided a good summary of the program goals (PNGV, 1997). In the committee's opinion, the PNGV has made substantial progress in assessing the potential of most candidate systems and identifying critical technologies that must be addressed to make each system viable. A few exceptions are noted in the sections describing specific technologies.

The committee has also described some international developments in the various technology areas, based both on its own knowledge and experience and on selected information gathering activities, but an extensive review of world-wide developments was not part of its task. Nevertheless, the issue of global competitiveness of the U.S. automotive industry is a key consideration in the development of advanced automotive technologies.

INTERNAL COMBUSTION RECIPROCATING ENGINES

The research team on the four-stroke direct injection (4SDI) engine has evaluated four engine configurations as candidate power plants: the CIDI engine, the homogeneous charge compression ignition engine, the gasoline direct injection (GDI) engine, and the homogeneous charge spark ignition engine. The PNGV has indicated that at this time the CIDI engine has the potential for the highest fuel conversion efficiency. Furthermore, because of the increased penetration of automotive diesel engines into the European market and a technical and manufacturing maturity that falls within the PNGV program schedule, there is a high level of confidence in the assessments of future improvements in CIDI engine performance.

In addition to better fuel economy, the performance of the CIDI engine is superior to other engine types in terms of evaporative, cold start, and hydrocarbon and carbon monoxide (CO) emissions. However, there are still significant challenges facing the development of CIDI engines that can meet the PNGV targets. The challenges include reducing the emissions of nitrogen oxides (NO_x) and particulates, reducing the weight of the power plant, and reducing costs.

The CIDI engine is being considered as a possible stand-alone power plant, as well as part of either a series or parallel HEV. The trade-off of fuel economy and weight involved in adding energy conversion devices with a hybrid vehicle design must be carefully evaluated because an increase in vehicle weight results in a decrease in fuel economy.

Program Status and Progress

The 4SDI team was very active this past year. A five-year comprehensive plan was developed, and the technical developments required for each component of a CIDI engine for a PNGV vehicle were identified. Technologies that would enable an advanced CIDI engine to meet Goal 3 objectives would include four valves per cylinder; a common rail, electronically controlled fuel injection system; a variable-geometry turbocharger; exhaust-gas after-treatment for NO_x and particulates; electric actuators; and an aluminum block. It would be fueled with a very low-sulfur diesel or alternative fuel. This engine would be significantly different from current diesel engines, which typically have two valves per cylinder; rotary pump fuel injection systems; fixed-geometry turbochargers; oxidation catalysts in the exhaust; pneumatic actuators; cast-iron structures; and use conventional diesel fuel. Technical developments are expected to reduce NO_x

and particulate emissions; reduce noise vibration and harshness (NVH); improve power density; and improve fuel economy.

Last year, the 4SDI team identified five high-priority areas for research: lightweight engine architectures; dimethyl ether (DME) as an alternative fuel for CIDI engines; combustion-related processes; lean NO_x catalysis; and alternative fuels for CIDI engines. The 4SDI team has been active in all five areas in the last year.

Lightweight engine structures are being investigated to reduce vehicle weight. Ford, for example, is testing the DIATA (direct-injection, aluminum-block, through-bolt assembly) engine, a 1.2-liter displacement engine designed to produce 45 kW/l. The design is a lightweight engine that achieves state of the art NVH.

Under a contract with the U.S. Department of Defense Tank Automotive Command, Ricardo, Inc., has developed a preliminary design for an all new, three-cylinder, lightweight, high speed, direct injection engine. The main objective of this program was to establish an engine architecture compatible with lightweight materials. The key challenge is placing the lightweight materials under compression, and through-bolt assembly was considered the most promising way to accomplish this. Key aspects of this engine architecture are expected to be used in the Chrysler-U.S. Department of Energy HEV program.

One of the critical challenges to the CIDI engine is the so-called trade-off between emissions of soot (particulates) and NO_x. In current diesel engines, methods used to reduce NO_x (typically increased exhaust-gas recirculation and retarded injection timing) result in increased soot emissions and vise versa. It is usual to display the diesel emission characteristics on a graph of soot versus NO_x. Obtaining a net benefit in emissions requires decreasing overall emissions toward the origin of the operating curve, rather than along the soot-NO_x trade-off curve. The CIDI engine technology under development is targeted to achieve a 0.04 g/mile particulate emissions level or better by (1) limiting the application of engine controls that reduce NO_x (e.g., exhaust gas recirculation) in order to minimize energy-out particulate emissions and (2) lowering NO_x emissions to target levels using catalytic after-treatment. There is some leeway in implementing in-cylinder NO_x reduction strategies, at the expense of increasing particulate emissions, while still meeting the total emission design target. Meeting the stretch research objective of 0.01 g/mile particulate emissions will require simultaneous reductions of soot and NO_x and cannot be met by manipulating the soot-NO_x trade-off relationship. Therefore, breakthrough improvements in engine controls to reduce emissions and for exhaust-gas after-treatment for both NO_x and particulates will be required, as well as significant changes in fuels.

The research objective for particulate emissions, therefore, will require fundamental investigations of the in-cylinder combustion process with the objective of altering the interaction between soot and NO_x emissions. To this end, PNGV has begun research on combustion fundamentals, such as combustion control through electronic control of the fuel-injection process and assessing the interaction between the combustion chamber geometry, the fuel injection, and fuel-air mixing. Ford and FEV have also been pursuing new technologies in fuel injection rate-shaping using piezoelectric techniques.

Many collaborative programs have been put in place in the past year. The 4SDI technical team participated in Vice President Gore's Technical Symposium Number 6 on 4SDI engines, which consisted of five sessions held over two days in the summer of 1997. The PNGV has established a Fuels Working Group and an Aftertreatment Working Group to develop PNGV strategy and plans. Cross-cutting teams were established to promote interchanges between the light-duty CIDI engine researchers and the heavy-duty diesel engine industry. For example, Chrysler is working cooperatively with Detroit Diesel Corporation to integrate a three-cylinder, 1.5-liter displacement, direct-injection, turbocharged, intercooled engine with hot and cold exhaust gas recirculation into an HEV. A DME fuel system has been designed and a follow-on program established.

Efforts to develop combustion systems are being augmented by experimental work at Sandia National Laboratories (SNL) and Wayne State University with computational support from the University of Wisconsin, Madison. SNL, Oak Ridge National Laboratory (ORNL), and Los Alamos National Laboratory (LANL) are pursuing improved lean NO_x catalysts, and Pacific Northwest National Laboratory (PNL) and Lawrence Livermore National Laboratory (LLNL) are investigating plasma-assisted NO_x reduction catalysis. A reformulated diesel fuel testing program is under way at EPA, and a USCAR-supported auto/energy fuel testing plan is being developed.

Technical Targets

The critical characteristics of a CIDI engine that can meet the PNGV performance targets are shown as a function of milestone targets in Table 2-1. All of the USCAR partners have made good progress towards meeting these targets.

A comprehensive evaluation of the Chrysler Generation I 1.46-liter, three-cylinder engine has revealed general conformance with the 1997 targets with respect to part load brake thermal efficiency, exhaust emissions of NO_x and particulates, based on a 14-mode test protocol. The one-meter noise assessment is in conformance with the 1997 target. Peak thermal efficiency is 2.5 percentage points below the 1997 target; however, improvements are expected in both the Generation I and II versions. Both displacement and weight-specific power are below target for 1997 by 5 and 13 percent, respectively. The latter shortfalls will be addressed by the Generation II design, which will incorporate more lightweight materials than the Generation I version. Mount vibration measurements have not yet been made.

Initial engine dynamometer tests have been made on the Ford Research 1.2-liter, four-cylinder engine. Displacement-specific power and engine noise results compare favorably to the 1997 PNGV targets. The part-load thermal

TABLE 2-1   Critical Characteristics of the CIDI Engine vs. PNGV Milestone Targets

Characteristic	Units	1995 Target	1997 Target	2000 Target	2004 Target
Best brake thermal efficiency	%	41.5	43	44	45
Displacement specific power	kW/L	35	40	42	45
Power specific weight	kW/kg	0.50	0.53	0.59	0.63
Cost per kW	$/kW	30	30	30	30
Durability	1,000 miles	150	150	150	150
NVH (one meter noise)	dBA	100	97	94	90
Engine-out NO_x emissions[a]	g/kW-hr	3.4	2.7	2.0	1.4
Engine-out particulates[a]	g/kW-hr	0.3	0.25	0.20	0.15
FTP 75 NO_x emissions in 2,500 lb ETW vehicle	g/mile	0.6	0.4	0.3	0.2
FTP 75 particulate emissions in 2,500 lb ETW vehicle	g/mile	0.08	0.06	0.04	0.04[b]

Source: Based on Table III.F-1 in PNGV (1997).

Acronyms: NVH = noise, vibration, and harshness; FTP = federal test procedure; ETW = emissions test weight.

   [a]Representative values for operation over the FTP cycle
   [b]In 1997, PNGV identified a stretch research objective for particulate emissions of 0.01 g/mile.

efficiency and emissions levels are very calibration-specific and are under development, making comparisons with existing engines difficult. Peak thermal efficiency is below the 1997 target but is also under development. The weight-specific power and package volume PNGV targets were established assuming a three-cylinder engine. This four-cylinder engine is still below the weight-specific power targets and the package volume targets.

Based on testing of GM's single-cylinder CIDI research engine, projected emissions data meet the 1997 PNGV targets. Other targets cannot be assessed from tests on a single-cylinder engine.

Because of the soot-NO_x trade-off inherent in CIDI engines, the more restrictive reserach objective for particulate emissions would alter the basis on which the CIDI engine has been evaluated as a potential PNGV power plant. To achieve the stretch objective, technological breakthroughs will be necessary for the CIDI engine to meet the PNGV milestones. A consequence of the more ambitious 0.01 g/mile research objective for particulate emissions is that fuel characteristics are now more important for meeting the PNGV goals. For example, the 0.01 g/mile particulate research objective corresponds to an emission of sulfates for a fuel with approximately 50 ppm of sulfur and a vehicle with a fuel economy of 80 mpg. Therefore, at a minimum, sulfur levels in the fuel will have to be drastically reduced from the current limit of approximately 500 ppm to about 50 ppm.

The stretch objective would make the CIDI engine a high-risk candidate for meeting the PNGV goals.

Meeting the stretch research objective for particulate emissions increases the overall challenge of meeting the exhaust emission standards. Reducing NO_x emissions is still one of the biggest challenges for the CIDI engine. Engine-out emissions appear to be 0.5 g/mile or greater, whereas the current emission target is the Tier 2 federal NO_x limit of 0.2 g/mile. Meeting these limits will require after-treatment of NO_x, with a NO_x conversion efficiency of 60 to 90 percent. Demonstrated after-treatment efficiencies are currently less than 40 percent, and many technologies under development require near sulfur-free fuel. Clearly, a high-efficiency, sulfur-tolerant after-treatment device for NO_x must be developed for the CIDI engine to be a viable option.

In addition to the higher priority of fuels technology, which should include investigating alternative fuels, such as DME and Fischer-Tropsch diesel fuel, the development of exhaust-gas after-treatment technologies must also be expanded to include methods for reducing particulate emissions.

Current Program Elements

Advances in engine combustion, exhaust-gas after-treatment and fuels technology will be necessary to meet the stringent PNGV emission requirements. The current program includes work on some aspects of all three of these technologies.

High-Pressure Fuel Injection Systems and Combustion Fundamentals

Significant advancements in combustion control will be necessary to meet the low emission targets without sacrificing fuel economy. The in-house research programs of all the USCAR partners on the fundamentals of combustion and emissions formation are being augmented by work at the national laboratories and universities.

The use of electronically controlled, high-pressure fuel injection systems as a means of combustion control is being investigated by the heavy-duty diesel engine industry. Next-generation electronic fuel injectors may allow for dynamic rate-of-injection profiling, in addition to multiple injections per cycle. NO_x, particulates, and fuel economy can be significantly affected in heavy-duty diesel engines by manipulating fuel injection. The extent to which these advanced technologies can be used to improve combustion in small CIDI engines, such as those that would be used in a PNGV concept car, is not known. The smaller CIDI engine will be operating at a higher speed than the typical heavy-duty diesel engine, the quantity of fuel injected will be smaller, and the combustion chamber will be smaller, so that surfaces will be closer together and fuel jets from the injectors will impinge on those surfaces; the injector holes, however, will be approximately the same size. As a consequence, the smaller high-speed CIDI engines have a shorter injection duration and less time available for mixing, so that controlling combustion through injection manipulation is uncertain. This is a serious technical challenge for the 4SDI team. Interaction between the PNGV and the heavy-duty diesel engine industry via the crosscutting team is an appropriate way for PNGV to address this issue.

Exhaust-Gas After-treatment

Exhaust-gas after-treatment represents one of the most challenging aspects of the 4SDI program. The treatment of exhaust gas, both of NO_x and of particulates, will be required to meet the program goals. Cooperative programs with Argonne National Laboratory (ANL), SNL, LANL, LLNL, ORNL, and PNL are already established. Both lean-NO_x catalysis and plasma-assisted after-treatment approaches are being investigated. Yet progress in this area has been slow. In addition, the performance of those technologies at the present state of development is adversely affected by sulfur in the fuel. The best "full brick" catalyst[1] with diesel engine exhaust, with diesel fuel added as a reductant, reduces NO_x by up to 37 percent at steady state over a narrow temperature range. Other reductants provide a higher efficiency but would require an auxiliary source of reductants onboard the vehicle. The fuel economy penalty of using a reductant is approximately 1 percent.

Plasma-assisted catalysis, a process in which an electrical potential difference generates nitrogen ions that combine with NO to form molecular nitrogen and atomic oxygen, shows promise of reducing both NO_x and particulate emissions. The current state of the art of plasma exhaust treatment technology requires that a catalyst also be used to maximize the reduction of the NO_x emissions. The results of early laboratory tests have been encouraging, but the technology must still be demonstrated on an engine. Estimates of the fuel economy penalty for plasma systems are in the range of 2 to 5 percent.

Particulate and NO_x traps are also being considered. Issues of cost, durability, and regeneration capacity remain for particulate traps, and the engine control systems for momentary fuel enrichment to release the NO_x from the trap and subsequent catalytic reduction are complex challenges that must still be met.

Despite the challenges, the emphasis on exhaust-gas after-treatment of NO_x and particulate matter will continue. Breakthroughs will be necessary for the development of a sulfur-tolerant, long-life, effective, passive NO_x removal device. In September 1997, PNGV representatives met with four major catalyst suppliers to discuss cooperative development.

---

[1]In the vast majority of laboratory catalyst tests, the catalyst is in the form of a powder. However, in current commercial catalytic converters, most of the catalysts are in the form of a coating on a monolithic structure. "Full brick" means that the catalyst test was conducted with a monolithic structure.

Fuels Technology

The relationship between the physical and chemical characteristics of a fuel and the emissions from an engine is the basis for government regulations on fuel properties. In diesel fuel, for example, the cetane number, aromatic content, and sulfur levels are all subject to regulation. However, the relationships between the chemical characteristics of a fuel and its physical properties, such as viscosity, lubricity, and cetane number, are extremely complex. For example, the cetane number of a fuel can be increased either by decreasing the aromatic content in favor of longer-chain paraffins or by adding cetane improvers. Both fuels will have similar overall combustion and emissions characteristics as assessed by today's metrics. In an engine designed to meet PNGV goals, slight differences in the composition and properties of fuels may also have a significant effect on the performance of after-treatment devices, such as NO_x catalysts.

Fuels containing oxygen generally produce less soot, which might be a basis for reducing emissions. However, the incremental costs and the effects on the infrastructure of this change in fuel composition have to be considered. Certain components in fuel, such as sulfur, can affect both combustion and exhaust-gas after-treatment systems. Sulfur in the fuel can contribute to soot emissions by forming sulfates; sulfur can also deactivate the exhaust catalyst.

Indeed the PNGV recognizes the importance of the interactions between fuel and engine performance and is involved in programs to reduce emissions through fuel changes. These programs are all based on the target of 0.04 g/mile for particulate emissions. The stretch research objective of 0.01 g/mile requires PNGV to change its research goals accordingly and concentrate on alternative fuels. PNGV should establish a cooperative program with the U.S. transportation fuels industry (see Chapter 5).

The PNGV already has some programs in place to evaluate alternative fuels. Under a contract issued in cooperation with the of U.S. Department of Defense Tank Automotive Command, AVL List GmbH conducted an assessment of DME as an alternative fuel for diesel engines, and DME is a candidate for further investigation. Because the physical characteristics of DME are much like those of propane, significant infrastructure changes would have to be made if DME were chosen. EPA is also evaluating reformulated and alternative fuels. The committee believes that all of these programs should be continued and that all alternative fuels should be investigated.

However, it appears that none of the operating regimes for any of the candidate engines will meet the design targets with the research objective (0.01 g/mile) for particulate emissions. Therefore, the potential for altering the soot-NO_x trade-off by manipulating the fuel formulation and the subsequent impact of fuel composition on exhaust-gas after-treatment devices is a critical issue for the 4SDI program, which must now investigate the engine and fuel as an integrated system. The fuels industry should be involved in this important area of development and research, just as it is in the European Auto Oil program and the Japan Clean Air Program (see Chapter 5 and Jones, 1997).

International Developments

Automotive Diesel Engines

The small direct-injection diesel engine is widely used in automobiles in Europe. Currently about half of all new cars sold in Europe are diesel powered, and it is estimated that as much as 30 percent of the entire fleet could be diesel powered by the year 2000. This market shift is being motivated by a combination of tax policies favoring the use of diesel fuel over gasoline, the high cost of fuel, and the consequent consumer demand for fuel-efficient vehicles. Diesel-powered automobiles are not as prevalent in Japan as they are in Europe, but they have a higher market penetration than in the United States, where less than 1 percent of the automobiles are diesel powered. Because there is little market demand for automotive diesel engines in the United States, domestic industries have little incentive to pursue critical technologies in this area. Therefore, it is not surprising that technical leadership in the critical areas of fuel injection and electronic control is outside the United States. The world leaders in automotive diesel engine injection and control technologies are probably Bosch (in Germany), Denso (in Japan), and Lucas CAV (in the United Kingdom). However, some important work is being done in the United States by Caterpillar and Navistar on automotive applications of electronically controlled, common rail, hydraulically amplified injection systems.

The partners in PNGV are well aware of the advancing state of the art in automotive diesel engines. In fact, through their foreign affiliates, they are participating in the development of these engines. GM has developed the Ecotec engine, and Ford has developed the DIATA engine in their respective European operations. Chrysler is involved in developing a state-of-the-art automotive-size diesel engine through working agreements with both a domestic and an international company. Although the United States cannot claim technical leadership in the general area of automotive-size diesel engines, the PNGV is well aware of the current state of the art and directions in development of this power plant. The USCAR partners are fully capable of utilizing this technology worldwide through their foreign affiliates and international agreements.

Gasoline Direct Injection Engines

In Japan, Mitsubishi, Toyota, and Nissan have introduced GDI engines to their domestic markets, claiming fuel efficiency improvements of 20 to 30 percent over the conventional spark-ignition engine vehicles. The PNGV partners have been following developments closely but have concluded that GDI engines would have a lower fuel conversion efficiency than CIDI engines and, like CIDI engines, they would not meet the U.S. emission standards (see Appendix B).[2] The belief that the maximum fuel economy of the GDI engine would be less than that of a CIDI engine was reported to the committee during the Phase 3 review by Dr. Peter Herzog of AVL (Herzog, 1996). Hence, the GDI spark-ignition engine was not listed as a candidate power plant for the PNGV concept vehicles.

As part of the Phase 4 review, Dr. Ando of Mitsubishi Motors presented impressive results of recent developments on their GDI engines (Ando, 1997 a,b; Iwamoto et al., 1997; Kume et al., 1996). Mitshubishi claims that significant improvements have been made in the total performance of the engine by changing the intake manifold design, altering the in-cylinder flow pattern, maximizing the distance between the fuel injector and the spark plug, carefully matching the fuel injector characteristics to the cylinder flow at different loads, and taking full advantage of the capabilities of advanced electronic controls. At this time Mitsubishi Motors believes the new design of the GDI engines will be able to meet the stringent European and U.S. low-emission vehicle (LEV) standards with fuel conversion efficiencies within 1 percent of CIDI engines. Mitsubishi estimates that by the year 2000 85 percent of the engines they produce will be GDI engines.

The advancements of the GDI engine claimed by Mitsubishi represent technical strides for this power plant. If these claims of improved performance can be realized, the GDI engine would be a viable competitor to the CIDI engine. However, even if the GDI engine meets the LEV standards at a fuel conversion efficiency within 1 percent of the CIDI engine, it is not known if it will meet the PNGV emission targets, which are the ultra low emission vehicle (ULEV) standards. The PNGV partners are aware of the GDI programs in Japan and are assessing the potential of the GDI as a power plant for a PNGV vehicle. The committee feels that this assessment should continue.

Assessment of the Program

Excellent progress has been made in the past year in all aspects of the 4SDI program. However, the identification of a stretch research objective for particulate emissions of 0.01 g/mile presents significant additional challenges to the 4SDI program in developing the CIDI engine as a PNGV power plant. The prospect of developing a CIDI engine that can meet this research objective is high risk and would make a reevaluation of other candidate engines and system configurations necessary. To maximize the probability of success, the 4SDI program may have to be augmented and redirected. The 4SDI team must determine if the new stretch research objective can be met by operating the engine in previously

---

[2]This view is detailed in Figure III.F-1 of the 1997 update of the PNGV Technical Roadmap (PNGV, 1997).

unattainable regimes or through new fuel formulations or alternative fuels. Cooperative programs with the transportation fuels industry would be a logical way to address this new challenge.

Recommendations

Recommendation. The PNGV should devote considerably more effort and resources to exhaust-gas after-treatment of NO_x and particulates. PNGV should consider greatly expanding its efforts to involve catalyst manufacturers.

Recommendation. A broad cooperative effort between the PNGV and the transportation fuels industry should be established to assess the potential for enhancing the performance of CIDI engines through fuel reformulation or alternative fuels. The PNGV should also assess the effects of fuel changes on the fuel production and distribution infrastructure.

Recommendation. In light of the published improvements in gasoline direct-injection engines, it would be prudent for the PNGV partners to continue to assess developments in this technology against PNGV targets and the CIDI engine, whether or not the gasoline direct-injection engine is chosen as a potential PNGV power plant.

CONTINUOUS COMBUSTION ENGINES

As the committee anticipated, continuous combustion engines—gas turbines and Stirling cycle engines—have not reached a state of development suitable for use in the year 2000 concept vehicles. Therefore, they now fall into the category of post-PNGV technology development. The committee concurs with this PNGV decision. However, continuous combustion, which can be controlled more easily than intermittent combustion, may require reevaluation if continuous combustion engines can be shown to meet the 0.01 g/mile stretch research objective more easily than internal combustion reciprocating engines.

Gas Turbines

The requirements established by the PNGV for an automotive gas turbine (AGT) engine in a series hybrid vehicle include a power level of approximately 50 kW, a thermal efficiency of 40 percent, a factory cost of $1,500 ($30/kW), and a weight of 60 kg (0.8 kW/kg). Because no candidate AGT engines approach the thermal efficiency goal, and in light of progress made with other candidate power plants, the PNGV has dropped the gas turbine engine from its list of promising technologies in 1997. As a consequence, the committee understands that the U.S. Department of Energy (DOE) will reduce support for research on AGT engines in fiscal years 1998 and 1999.

Program Status and Progress

The development of AGT engines has been supported by the government and private sector since the 1950s. To achieve high thermal efficiency and the necessary high operating temperatures, a number of development programs have focused on ceramic AGTs. Programs funded, or partly funded, by the DOE to develop ceramic AGTs have been conducted primarily by GM/Allison and Allied Signal with major contributions by ceramic suppliers. In the past decade, progress has been made toward achieving the desired properties and quality of the ceramic materials. Improvements have also been made in critical technologies associated with air bearings, low-emission ceramic combustors,[3] ceramic rotary regenerators/ low-leakage seals, ceramic axial and radial turbine rotors, ceramic stator/static structures, and insulation systems. By the end of 1996, component tests had been run on ceramic turbine rotors, combustors, and regenerators, but no fully inte-grated AGT engine with high performance had been demonstrated, and the long-term reliability of the ceramic components had not been proven.

AlliedSignal has focused on placing ceramic stator vanes into existing products—notably airborne auxiliary power units and military ground carts— and has undertaken the task of bringing a new material system into production in these applications. Allison has continued its efforts to demonstrate operation at turbine temperatures of 1,370°C (2,500°F) on turbine stages that can withstand abusive operating cycles and impacts by foreign objects. In 1997, the DOE-spon-sored programs were reduced in scope and now emphasize only the development of ceramic components. No new DOE-sponsored automotive ceramic AGT pro-gram contracts are planned after 1997. Phaseouts of existing contracts will con-tinue into fiscal year 1998 and possibly fiscal year 1999. Development of gas turbines for stationary power applications will continue in other DOE programs.

Although most AGT development programs had been focused on free tur-bine, prime propulsion engines, in the past few years interest has shifted to turbogenerator use in HEVs. Metal turbogenerators have been purchased by the auto companies for use in the DOE HEV programs and other hybrid demonstra-tions, including a 30-kW engine designed by Capstone Turbines, a 60-kW engine designed by AlliedSignal, and a 40-kW engine designed by Williams Research. With a best thermal efficiency potential of only about 32 percent, these engines have provided useful HEV demonstrations, but no production program is ex-pected. With the metal AGT as an energy converter, there is no real potential for achieving the PNGV fuel economy target of 80 mpg.

Programs to develop ceramic turbogenerators were briefly initiated with Allison (under GM's HEV Program) and Teledyne (under contract to Ford), but

---

[3]Two main approaches to ceramic combustors are the rich/quench/lean combustor, which is de-signed to avoid stoichiometric fuel-air conditions and associated temperatures to reduce the formation of NO_x, and the catalytic combustor, which uses catalysts to promote combustion at reduced tempera-tures and consequently reduce the formation of NO_x.

both were canceled before they could produce meaningful results. Thus, no serious future effort or investment is being planned in the United States to produce an AGT that can meet the PNGV targets. Gas turbine manufacturers have indicated that they could develop small turbogenerators approaching 40 percent efficiency, but development is unlikely because the estimated development cost is about $75 million. Even with this expenditure, there would be substantial technical risk with regard to the cost and durability of ceramic components, as well as uncertainty about the market.

Cost Issues

Informal discussions between committee members and gas turbine experts indicated that small (50-kW class) AGTs could be manufactured in large quantities at a factory cost of around the target value of $1,500; however, the committee did not verify this claim. The greatest uncertainty pertains to the cost of ceramic components, which are based on low-cost raw materials and do not have the high costs of superalloys but are unproven in terms of net shape process yields, forming, and inspection costs. Because the gas turbine is radically different from the current reciprocating engines, massive infrastructure costs would be entailed— both in manufacturing plants and in maintenance training and facilities—if gas turbines were to be adopted into production automobiles.

International Developments

In 1997, Japanese programs reportedly demonstrated ceramic turbo-generators, one of which was directed toward automotive use and demonstrated 32.3 percent thermal efficiency at 92 kW (Nakazawa et al., 1997). A second, which was directed toward industrial and truck/bus applications, demonstrated 37 percent thermal efficiency at 240 kW (Ichikawa et al., 1997). Both engines were operated for more than 100 hours with ceramic components, including regenerators, combustors, turbine rotors, and static structure.

Assessment of the Program

Program Direction. If internal combustion engines can ultimately achieve close to 45 percent thermal efficiency and can economically meet the stringent emission regulations, or if fuel cells using gasoline reformate can be developed with similar efficiencies and lower or equal system costs, then the development of ceramic gas turbines by PNGV should be terminated. If not, the AGT, which has low emissions,[4] the potential to provide a thermal efficiency approaching

---

[4]NO_x problems become worse at higher temperatures. At the temperatures necessary to achieve high efficiencies in gas turbines, NO_x generation could be a significant problem.

40 percent, and small size and weight, should be re-evaluated. Considering the risks inherent in developing a CIDI engine or a fuel cell that can meet all requirements, it may be premature to eliminate all PNGV-sponsored AGT technology development.

Ceramic Components. In spite of significant progress and successful production and durability testing of individual ceramic components, they have not been demonstrated to the point of acceptable risk for the development of a ceramic gas turbine for automotive application.

Recommendation

Recommendation. In view of the stringent emissions targets, the PNGV and DOE should continue to support the development of ceramic component manufacturing and durability demonstrations on a limited basis.

Stirling Engines

Although the Stirling-cycle engine has not been within the scope of the PNGV/USCAR joint activities, GM has a program to install a 30-kW engine in a series hybrid vehicle as part of DOE's HEV program. The engine technology is proprietary to Stirling Thermal Motors of Ann Arbor, Michigan. It is believed that a thermal efficiency of about 30 percent has been achieved by this automotive power plant and that many technical problems have been overcome. One exception may be the long-term containment and retention of the hydrogen working fluid used in this closed-cycle engine. Emission levels and other performance data have not been published.

GM has projected a potential for 36 percent thermal efficiency for the Stirling-cycle engine, which would fall somewhat short of the performance projections for other energy converters. Most observers believe that the continuous flow, external combustor has the potential for excellent emission levels, a principal advantage of the Stirling engine; however, the large size and complexity of this engine, which imply higher cost, are considered competitive disadvantages. The current demonstration under DOE's HEV contract has essentially been concluded.

Recommendation

Recommendation. The PNGV should continue to monitor the nonautomotive development of Stirling-cycle engines but should consider further development only if warranted by NO_x and particulate emissions considerations.

FUEL CELLS

Of all of the technologies being considered in the PNGV program to convert fuel energy into useful power, fuel cells still offer the best long-term potential for high efficiency and low emissions.[5] A fuel-cell vehicle does not have to be a hybrid system if gaseous hydrogen is the fuel. However, most liquid hydrocarbon-fueled fuel cell systems are hybrid systems because they have faster startups (initial driving can be powered by a battery), better transient operation (a battery can be used to augment fuel cell power during rapid demand changes), and regenerative energy recovery. All of the PNGV fuel cell systems presented to the committee thus far have been hybrid systems. However, in spite of considerable, even impressive, progress, fuel cells still face substantial obstacles to reaching performance and cost goals in the PNGV 2000 to 2004 time frame.

It has become very clear that the development of a successful automotive fuel-cell system is intimately connected to the choice of fuel. Hydrogen, which combines electrochemically with oxygen to produce electric energy, namely, an electrical current and a potential (voltage), is the only part of the fuel that is utilized in the fuel-cell stack. However, pure hydrogen, which is an excellent fuel for fuel cells, is much more difficult to store onboard a vehicle than liquid hydrogen-containing hydrocarbon fuels. Vehicle studies have shown that it is very difficult to store enough compressed hydrogen gas on board a PNGV-type vehicle to travel more than 100 miles. Furthermore, even this range requires high pressures (3,000 to 6,000 psi) with correspondingly heavy (and expensive) storage tanks and the energy losses associated with compressing hydrogen. Transportation of hydrogen fuel from production plants is also expensive, and present projections show unit energy costs to be several times those of petroleum-based fuels. Furthermore, there is virtually no infrastructure for making hydrogen gas available to consumers. Thus, until onboard storage, unit energy costs, and infrastructure problems are resolved, gaseous hydrogen fuel will probably be practical only for certain commercial and vehicle fleets.

Most of the performance obstacles and many of the cost hurdles of fuel-cell systems are associated with the necessity of storing and using liquid hydrocarbon fuel onboard the vehicle. In fact, the PNGV program seems to consider the use of gasoline as a basic requirement because of the virtual absence of infrastructure for alternative fuels (e.g., hydrogen, ethanol, and methanol). This requirement solves the near-term fuel infrastructure problem but aggravates the near-term performance efficiency and cost problems.

---

[5]A well designed and properly operating system integrating a reformer with a fuel cell should have virtually no emissions of oxides of nitrogen, hydrocarbons, or carbon monoxide, although this has not yet been demonstrated on automotive-compatible reformer systems. However, there may be emissions during cold start. After-treatment of exhaust emissions may be required, but the technology for removing hydrocarbons and carbon monoxide, once the catalyst is warm, would be simpler than three-way catalysts since no NO_x removal would be necessary.

In the past, the technical teams paid little attention to fuel issues, but fuel strategy has now become a major issue. One result of this change is that DME is now considered a fuel potentially suitable for both diesel engines and fuel cells. Until fuel cells can be introduced in large numbers, alternative fuels may be an attractive option. Before fuel cells can become prevalent, however, the difficulties and inefficiencies of processing fuels onboard the vehicle related to transients, small size units, and the temperature mismatch between gasoline reformation (about 700°C) and the proton-exchange-membrane (PEM) stack (less than 100°C) must be overcome. Furthermore, if the reduction or elimination of carbon dioxide (CO_2) emissions becomes a dominant consideration, off-board fuel processing to produce hydrogen and sequester CO_2 or carbon could be another factor in favor of alternative fuels.

Providing hydrogen from a hydrocarbon fuel like gasoline requires an onboard fuel processor, which adds weight, volume, and cost to the system. It also adds start-up delays, slows transient response times, and reduces fuel conversion efficiency. It decreases stack performance, complicates system integration, and has the potential to produce some (very low level) emissions. Indeed, many performance aspects of fuel-cell stacks would be adversely affected, including (1) cell voltage at a given current density (mA/cm^2), a measure of cell efficiency; (2) maximum values of current density for a given minimum voltage, a measure of cell surface area required to produce a given power; (3) catalyst loading of precious metal (mg/cm^2), an important cost factor; and (4) specific power (kW/kg), a measure of stack weight for a given design power. Not surprisingly, then, a substantial part of the DOE-sponsored PNGV fuel-cell development (about 30 percent) has been directed toward reformers.

Significant attention has also been directed toward making stacks more tolerant to CO (produced by reformers) and toward improving the post-reformer CO cleanup. The CO cleanup utilizes preferential oxidation techniques to reduce the amount of CO that goes from the reformer to the stack. In short, PNGV's efforts have been directed both toward reducing the amount of CO reaching the stack and increasing the amount of CO the stack is able to tolerate. Because heavy precious metal loadings are required to improve CO tolerance, efforts are also under way to find alloy catalysts, which are less expensive alternatives to pure platinum.

Program Status and Progress

Stacks and Stack Systems

Significant progress has been made in the development of fuel-cell stack systems. An International Fuel Cells 50 kW hydrogen-fueled PEM system was operated at near atmospheric pressure (1 to 2 psig, i.e., about 1.07 to 1.14 atm) with 50 percent PEM system efficiency (hydrogen-to-electricity conversion) at 100 percent power, and with 57 percent system efficiency at 25 percent power.

This system provided a specific power of about 0.37 kW/kġ. Up to this point, it had been generally assumed that it would be necessary to pressurize the stack to at least 2 or 3 atmospheres. Pressurization does improve stack performance, but it requires adding a compressor and expander to the system, with accompanying costs and complexities. Thus, if acceptable performance can be demonstrated with little or no pressurization, it could ease operational problems and reduce costs.

A Ballard 30 kW PEM system operating on methanol reformate was also demonstrated. It showed a steady-state efficiency (at constant output level, methanol fuel-to-electric [direct current] power conversion) of 42 to 44 percent and a CO tolerance of 40 ppm. The fuel processor was integrated with the stack.

A small-scale cell (50 cm^2) configuration with increased tolerance to CO (up to 100 ppm) and durability of at least 1,400 hours was demonstrated at LANL. However, the tests were conducted with hydrogen fuel (as opposed to reformate), and the catalyst loading was 0.9 mg/cm^2, far above levels compatible with PNGV cost goals. Additional life tests are planned using 40 percent hydrogen in a simulated reformate. Short-term tests of cell performance on partial oxidation reformate (40 percent hydrogen) showed about 15 percent performance degradation compared to pure hydrogen.

Reformers

Delphi has demonstrated a 30-kW methanol reformer integrated with a preferential oxidation (PrOx) system. This reformer provided about 20 ppm of CO at 100 percent load and 40 ppm at 25 percent load and a reported reformer system efficiency (methanol-to-hydrogen conversion) of 85 percent at 100 percent load.[6] The PrOx CO cleanup system accounted for about 20 percent of the system weight and volume.

One of the most visible accomplishments in the past year was the successful operation of a partial oxidation reformer, designed by AD Little, integrated with a PrOx system designed by LANL. Tests utilizing low-sulfur gasoline showed an output of about 40 percent hydrogen and 50 ppm of CO at an oxygen stoichiometry of about 1.25.[7] In another test, an AD Little Gen-2 50-kW fuel processor was integrated with a LANL three-stage PrOx CO-cleanup system. The resulting reformate (6 to 40 ppm of CO) was fed into Plug Power and Ballard PEM stacks (0.5 kW to 5 kW). Perhaps the most impressive part of this test was switching from ethanol fuel to gasoline while the system was in operation. This test appeared to demonstrate multifuel capability, although the hardware was far from

---

[6]Conversion efficiencies refer to a lower heating value of hydrogen fuel out relative to a lower heating value of hydrogen fuel in.

[7]An oxygen stoichiometry of 1.25 implies 25 percent more oxygen than necessary to oxidize all carbon to carbon dioxide.

acceptable for automotive applications. An issue that must still be addressed is the sensitivity of the catalytic system to the sulfur content in the fuel. Either a sulfur-removal device or a low-sulfur fuel will be necessary for durable operation.

Progress on fuel processors was also demonstrated with a "microscale" version of an autothermal reformer that showed high efficiency (about 87 to 93 percent) in producing hydrogen from methanol, ethanol, and gasoline. GM is currently validating this reformer for methanol for a 50-kW fuel processor.

Fuel Flexibility

Substantial improvements have been made in matching the CO level produced by the fuel processor (20 to 40 ppm) and the CO level that can be tolerated by the fuel-cell stack (perhaps 100 ppm with an air bleed), suggesting that the choice of fuels could be flexible; it may be possible to use gasoline, methanol, ethanol, or methane. However, a clear picture of the efficiency trade-offs is necessary to achieve this match. The overall projected efficiency (direct current electrical energy out relative to fuel energy in) of the fuel-cell system must also be compared with other technologies, including losses that might occur during fuel processing before the fuel gets to the vehicle (methanol, DME, hydrogen).

Gasoline is a blend of hydrocarbon fuels and small quantities of other substances, including sulfur. The relative amounts of the hydrocarbon fuels and the other substances vary across fuel manufacturers and their retail outlets, and composition can change over time. Thus, it may be difficult to optimize fuel processor performance, and the sulfur in gasoline could have adverse effects on catalysts in the fuel processor or PrOx system and in the stack.

Cost Issues

Cost projections for gasoline-fueled fuel-cell systems have decreased dramatically but are still very high at $500/kW, nearly double the 1997 goal of $300/kW and about an order of magnitude higher than the 2004 PNGV goal of $50/kW. However, for the first time, rigorous cost analyses are being conducted by Directed Technologies, Inc., for stacks, fuel processors, and ancillary systems. These analyses could help determine ways to lower costs for high-volume manufacture (e.g., manufacturing techniques, materials selection, etc.). Until low-cost membrane-electrode assemblies, low-cost bipolar plates, lower catalyst loadings, and low-cost reformer/cleanup systems are developed, the cost of reformers will remain a major issue.

International Developments

The most visible foreign developments are associated with Ballard Power Systems, Inc., of Canada, and Daimler Benz of Germany. Ballard is forming strategic partnerships and receiving orders for PEM systems from automobile manufacturers worldwide. Probably, the activities with the greatest potential for a long-term effect on automotive fuel-cell development are Ballard's $450 million (Canadian) joint venture with Daimler Benz and Daimler's apparent commitment to fuel-cell vehicles. Daimler introduced two prototype fuel-cell vehicles in 1997 and acquired a 25 percent stake in Ballard in the joint venture. One of the vehicles was a 250-kW PEM hydrogen-fueled bus; the other was a hydrogen-fueled mini-van, NECAR III. Daimler is also expected to introduce a prototype methanol-powered subcompact car based on the new A-class vehicle. Meanwhile, Ballard is supplying fuel cells to Delphi, Chrysler, and Nissan and has joined the Ford P2000 project. Ballard is also expected to deliver three more fuel-cell powered buses to Chicago and three to Vancouver for testing and evaluation. In addition, they are providing a methanol-fueled PEM system for the Georgetown University (in Washington, D.C.) 40-ft bus program. Ford has announced that it will join Ballard and Daimler Benz with an investment of $420 million in cash, technology, and assets in the development of fuel-cell technologies for vehicles (Ford, 1997).

A European program is under way to develop and test a 30-kW fuel-cell powered Peugeot van in partnership with PSA (a French company) and Deltona and Ansaldo (Italian companies). In Japan, Nissan and Mitsubishi are working on government-sponsored R&D, and Honda and Toyota appear to be making major corporate investments to develop their in-house capabilities. The combined government-industry investments in transportation fuel-cell technologies in Europe, Japan, and Canada are significant, and individual USCAR partners have responded with ambitious investments and programs of their own.

Assessment of the Program

Tests of fuel processors, stacks, and complete systems to date have not included emissions and energy efficiencies for cold start-ups, shutdowns, or even rapid transient operation. The same is true of tests of component models and system simulation studies.

Even though PNGV is committed to the development of systems that will utilize gasoline for fuel, the committee is not aware of any efforts by PNGV to determine the sulfur tolerance of the stack platinum catalyst or the catalysts in the fuel processor or PrOx cleanup system. A requirement for sulfur-free gasoline could be a challenge for gasoline manufacturers.

In general, research on automotive fuel-cell systems has focused on systems that require pressurization to 2 or 3 atmospheres, which will require a small, high-efficiency, low-cost compressor, as well as a companion expander (turbine), to recover part of the stack air discharge energy. Because the compressor for these pressurized systems could consume 25 percent (or more) of the stack gross power output, the importance of high efficiency is obvious. DOE currently has contracts for the development of compressors with three firms: AlliedSignal (turbo compressor-expander), AD Little (scroll compressor-expander), and Vairex (variable displacement compressor). All three contracts have completed Phase I and are going into Phase II. However, viable solutions to meet efficiency, performance, and cost requirements compatible with PNGV fuel-cell system goals have not been demonstrated.

Progress has been made in the development of models of fuel-cell systems, primarily by ANL and LANL, which have delivered fuel-cell/fuel-processor simulations to the PNGV systems analysis team. However, these models appear to be very modest compared to the models required for full performance-range simulations. In spite of the likelihood that effective modeling and simulation could dramatically improve the development of appropriate systems and subsystems, modeling is still a low priority for PNGV.

In spite of significant reductions of estimated cost for the stack and other subsystems, cost is still a major uncertainty for fuel-cell viability. The costs associated with several critical (and expensive) subsystems, such as the compressor, turbine, fuel processor, reformate cleanup, and control, are all but unknown.

Recommendations

Recommendation. The PNGV should place a very high priority on additional modeling and system analyses for fuel-cell systems. Analyses of emissions and the overall efficiency of gasoline-fueled systems over the full range of operations, from cold start-up to shutdown, should be done, as well as trade-off studies for alternative fuels, such as methanol versus gasoline.

Recommendation. The PNGV should attempt to determine the sulfur and carbon monoxide tolerance of the fuel-cell stack platinum catalyst and the catalysts in the fuel processor and preferential oxidation cleanup system.

Recommendation. The PNGV should expand and accelerate its cost studies to include critical items, such as compressor expanders, fuel-flexible reformers, and carbon monoxide cleanup systems.

Recommendation. Because of their high efficiency and low emission potential, fuel-cell systems for transportation systems could become vitally important to the United States. U.S. government and industry investments in research and development should, therefore, be continued at current levels or even increased for an extended period.

ELECTROCHEMICAL ENERGY STORAGE

HEVs use energy storage systems to recover and to store braking energy, thereby enhancing overall vehicle system efficiency. Energy storage devices also provide some power during transient and peak power periods, allowing a smaller engine to operate at more constant power, improving efficiency and reducing emissions. Among electrochemical energy storage technologies, batteries are at a considerably more advanced state of development than ultracapacitors. Because commercial lead-acid and nickel-cadmium rechargeable batteries do not meet the PNGV performance and cost criteria, attention has been focused on lithium-ion and nickel metal hydride batteries.

HEVs require a higher-value battery power-to-energy ratio than battery-powered electric vehicles because relatively little energy has to be stored onboard hybrid vehicles, as compared with electric vehicles. Two types of power plants have been assumed in the PNGV analysis, fast-response and slow-response power plants. A fast-response power plant, such as a reciprocating internal combustion engine, is capable of reaching its maximum power in a fraction of a second. Fast-response power plants place lower demands on the energy storage device than power plants that take several seconds to develop full power. For slow-response power plants (such as reformer fuel-cell systems), the energy storage device must deliver the difference between the power demanded for rapid vehicle acceleration and the power available from the power plant, and it must deliver this power for a longer period of time. This increases the peak power demand, as well as the energy required from the storage device. Table 2-2 shows the PNGV power, energy, and other design targets for short-term energy storage.

TABLE 2-2   PNGV Design Targets for Short-Term Energy Storage

Characteristic	Units	Fast Response	Slow Response
Discharge pulse power (18 s)	kW	25	65
Peak regenerative power (10 s)	kW	30	70
Available energy	kWh	0.3	3
Discharge power density	kW/L	0.78	1.6
Minimum round-trip efficiency on the FUDS/HWFTET cycle	%	90	95
Discharge-specific power	kW/kg	0.63	1.0
Cost	$/kW	12	7.7
Durability (100 Wh)	cycles	50,000	120,000
Lifetime	yr	10	10
Operating temperature	°C	−40 to +52	−40 to +52

Source: PNGV (1997).

Note: A fast-response power plant is assumed to react very much like a conventional automotive engine, which responds very quickly to vehicle power demands. A slow-response power plant puts a much greater demand on the energy storage system for the instantaneous delivery of high power. Each cycle is about one minute long and includes a discharge pulse during acceleration, a smaller discharge current at cruising speed, a strong charge pulse during braking, and a rest period. Superimposed on this is a charging current, which has the net effect of restoring the battery to the same state of charge at the end of the cycle as at the beginning.

FUDS = federal urban drive cycle; HWFTET = highway fuel economy test.

In the third report, the committee documented significant progress, especially in the development of high-power batteries (NRC, 1997). Since then, further progress has been made in lithium-ion and nickel metal hydride battery technology, and in all likelihood target design and performance goals can be approached. As a result, the year 2000 concept vehicle will not be unduly compromised. However, meeting the cost goals is still a major challenge.

Program Status and Progress

Lithium-Ion Batteries

SAFT is the principal contractor for the development of lithium-ion batteries for HEVs. In the past year, SAFT has designed and constructed 10 6-Ah cells. The specific energy of these cells has not reached the target, but the cycle life is close to the goals shown in Table 2-2. SAFT has also conducted some abuse-tolerance tests for 6-Ah cells, including nail penetration and mechanical shocks. Abuse-tolerance tests on 12-Ah cells were completed by the end of January 1998. Facilities for fabricating full-size cells (12 Ah) have been installed, and preliminary designs have been completed for the 50-volt module. The modeling and trade-off analyses are somewhat behind schedule; the results to date are insufficient to predict the optimized design, performance, and costs of the 50-volt modules. Costs remain significantly higher than the targets.

New exploratory projects have been initiated at PolyStor and VARTA to assess alternative baseline technologies. Both of these are six-month contracts that started in July 1997. A second program to develop 50-volt modules may be initiated at VARTA.

Nickel Metal Hydride Batteries

Nickel metal hydride batteries have reached the manufacturing stage for applications in electric vehicles commercialized by GM, Toyota, and Honda. VARTA, the principal contractor to DOE for development of a hybrid vehicle application, was able to meet its performance goals for 10-Ah nickel metal hydride cells. The specific energy is still below the target value of 60 Wh/kg with a power-to-energy ratio of 25 W/Wh. Tests of VARTA cells at Idaho National Engineering and Environmental Laboratory included determinations of (1) capacity and discharge/regenerative pulse power vs. rates at different temperatures, and (2) efficiency loss due to cycling at very low charge/discharge rates. Cycle life tests are in progress.

Like the modeling studies for lithium-ion batteries, modeling for nickel metal hydride batteries is in the preliminary stage, and the projected costs are high by a factor of about four. A Phase 2 program was initiated at VARTA in August 1997 to design, construct, and test a 50-volt module. The contract is for an 18-month period.

Ultracapacitors

The exploratory R&D on ultracapacitors has been completed. The technology is too immature for the PNGV time frame, although it is being considered for use in the control of the power-electronics system. The committee agrees with the PNGV assessment that breakthroughs will be necessary to increase the specific energy of ultracapacitors significantly and to lower the costs before they can be considered as primary HEV energy storage devices.

Assessment of the Program

Considerable progress has been made in the development of full-size cells of lithium-ion batteries and of nickel/metal hydride batteries. However, specific power and energy must still be improved, and the degradation in performance associated with cycling, particularly in the case of lithium-ion batteries, must be minimized.

PNGV should undertake fundamental investigations to elucidate the increased cell resistance and decreased capacity and power associated with cycling of lithium-ion batteries. For PNGV's purposes, the energy efficiency of the battery will have to remain high throughout the life of the battery. Test results should indicate energy efficiency, as well as the usual specific energy and power and cycle life.

PNGV recognizes the uncertainties of choosing the best battery and has opted instead to continue working on two main systems and monitoring progress on lead-acid and nickel-cadmium batteries. Of the two main systems, the lithium-ion system promises higher performance, but the nickel metal hydride system is more likely to meet the goals for 2000 and 2004. In general, the battery program is well directed, with proper emphasis on overcoming the obstacles to meeting the program goals.

Modeling analysis is still in its infancy and appears not to be coordinated well with modeling by the vehicle systems analysis team. Systems analysis (see Chapter 3) involves going back and forth between the requirements of the vehicle mission and the capabilities of the various subsystems and requires better models for the electrochemical energy storage systems so that realistic goals and targets can be set. The PNGV did not inform the committee of progress toward certain targets, such as the energy efficiency of the energy storage device, which is essential to a hybrid system achieving high mileage. For a better understanding, the committee would need to know the test protocols in more detail, as well as the test results.

Safety, particularly for the lithium-ion system and the 50-volt modules, is closely related to the need for efficient thermal management and electronic control of individual cells during cycling. The performance of the batteries and their safety over the temperature range indicated in Table 2-2 has not been reported.

Previous studies have suggested that performance, including cycle life and safety, is compromised at higher temperatures. Safety is generally considered a key issue even for small batteries in consumer electronics. Safety issues must be addressed and reported thoroughly in the PNGV program, which involves much larger batteries and fewer opportunities for controls on individual cells. Methods for controlling and monitoring cell level currents must be investigated further.

At the present time, the costs of all the battery systems under consideration are sufficiently high that it seems unlikely that the cost goals can be realized within the PNGV time frame. A detailed cost analysis with a complete breakdown would shed some light on ways to reduce costs.

International Developments

Several international battery developers (Matsushita-Panasonic, SONY, Japan Storage Battery, VARTA, and SAFT) are exploring high-power batteries for HEV applications. Matsushita-Panasonic is developing a 20-Ah capacity battery. SONY uses lithium cobalt oxide for the positive electrode material. Japan Storage Battery is investigating an alternative material, lithium manganese oxide. Honda and Mazda plan to utilize an ultracapacitor for regenerative braking in an HEV. The committee was encouraged that the U.S. PNGV contractors appear to be well ahead in energy storage technology for HEV applications.

Recommendations

Recommendation. The PNGV should expand its programs to include safety issues, such as temperature limits for lithium-ion batteries and preventing the generation of potentially explosive gases (e.g., hydrogen) in nickel metal hydride batteries.

Recommendation. The PNGV should conduct a detailed cost analysis to identify major contributors to high cost and establish strategies for reaching the cost goals.

Recommendation. The PNGV should update the storage requirements and goals by means of subsystem models integrated with the overall system analysis. In addition to specific energy, test results should be reported on energy efficiency and specific power over well defined test protocols and compared to the refined goals.

FLYWHEELS

Because of their attractive power-to-weight and power-to-volume characteristics, flywheels continue to be considered for energy capture and delivery in support of the efficiency of the "core power system" for PNGV vehicles. Flywheels excel in accepting the high power generated during engine and vehicle braking and delivering it to meet vehicle system peak power needs. The issues of safety, cost, and size are still serious but are yielding to development programs.

Work continues at ORNL to develop more specific flywheel design guidelines to support fast-response power-plant systems. A fast-response power plant, which has a similar response time to a conventional engine, places a much lower requirement on the power output of the flywheel than a slow-response power plant. A vehicle system model is now available that can be used for simulations to optimize the performance of the flywheel and to update the PNGV technical targets, which have been unchanged since 1996 (see Table 2-2). Flywheel designs will not be pursued for slow-response power plants because the much greater energy demands would require a larger and more costly flywheel system.

Program Status, Progress, and Plans

The PNGV flywheel technical team is now confident that it is possible to design and build a practical prototype energy storage flywheel system for automotive applications. A significant amount of work on failure containment has provided more confidence that the system can be designed to comply with the established safety criteria. Furthermore, improved containment designs have reduced cost and weight, which lends further support to the expectation that a practical system can be designed.

A key technology development is the design of an adequate containment mechanism in case of failure. The flywheel technical team has followed several strategies and has essentially overcome this significant barrier. Perhaps the most important advance is the growing evidence that flywheels (or portions thereof) that fail at low stress-to-strength ratios do not "burst" but remain intact. This knowledge dictates that the rotating parts have a high ultimate strength-to-maximum operating stress ratio (about 4:1).

Retaining "loose flywheels" is significantly easier than containing fragments because of the increased time for energy dissipation. The new design strategy for flywheel housings are designed to retain loose flywheels and contain fragments from partial flywheel failures instead of containing a complete burst and disintegration of a flywheel. This design strategy also attempts to manage energy as it dissipates. Limiting the use of flywheels to fast-response power plants reduces the energy storage requirement and permits the design to meet the safety goal for strength-to-stress ratio with a manageable increase in weight.

A variety of flywheel containment tests have been run in the Trinity/LLNL project, which led to a lightweight stainless steel/aluminum honeycomb structure that has so far been tested 41 times in sample form and has demonstrated consistent retention of test projectiles.

The current overall state of development for a flywheel system with a power level of 30 kW and an energy storage of 300 Wh indicates that, although the projected cost and weight are still well above PNGV targets, substantial improvements have been made in the last year. The assumptions in the projections include a cost of $5/lb for carbon fiber material, a containment-to-rotor weight ratio of 2:1, and a cost of $4/kW for power electronics. The first two projections appear to be achievable by 2004, but the $4/kW target is $3/kW lower than the $7/kW target set for the power electronics overall, which is already considered an extremely ambitious cost target. The relative simplicity of the flywheel electronic controls might justify these lower cost estimates and target, but the committee is not convinced of this. The cost for the flywheel, including the containment housing and motor generator, is substantially higher than the target; the weight is also significantly over its target. The amount of effort that will be required to achieve the desired targets is clearly much less than was projected a year ago but is still sizable. The flywheel technical team noted that if the power plant system required a flywheel with a capability of 15 kW power and 200 Wh energy-storage, current estimates of cost and weight would practically meet the targets.

A failure mode and effects analysis for the vehicle system incorporating a flywheel was conducted at ORNL, and a flywheel simulation model has been provided to the systems analysis team by LLNL. Assumptions for the current flywheel system are that life-cycle tests will demonstrate that the vacuum in the flywheel chamber will be maintained and that the sealed bearings will continue to operate with acceptable characteristics. The flywheel technical team and ORNL have yet to confirm the basic shape of the flywheel system, which may have implications for gyroscopic effects and containment costs.

The development of analytical models will continue at ORNL and LLNL in 1998 to support design assumptions and containment design and development. The flywheel design assumptions will be written to support performance of a design level failure model and effects analysis, which in turn will be used to support requests for quotes for the flywheel system in 1999. With further refinements of the vehicle system model with iterations of flywheel system capabilities, estimates for the targets, which are currently considered rough projections, will be improved. The Trinity/LLNL project will continue testing the containment with a full-size housing subjected to overspeed burst.

Assessment of the Program

The flywheel technical team is confident that it is now possible to design and build practical energy storage flywheel systems for automotive applications. The team, working with a number of outside agencies, all of whom are providing corroborating data of various kinds, has a firm grasp on the key technology barrier, namely, containment of failure. The committee suggests, however, that an out-of-balance sensor be considered to shut off electrical power during flywheel run-up if a significant amount of out-of-balance vibration is detected. The committee believes the flywheel system should be included in second-generation concept vehicles to help clarify how the vehicle system and power plant will incorporate the energy captured by the flywheel and enable the development of a delivery system that optimizes overall cost, size, performance, and efficiency.

Lowering the cost and reducing the weight of the flywheel system are still necessary, but with a smaller flywheel system, the tasks will be easier to accomplish because the power and energy storage requirements will be lower. The committee believes that the relatively large cost penalty of $300 for the flywheel system will be extremely difficult to offset by reductions elsewhere in the vehicle system.

Recommendation

Recommendation. If vehicle systems modeling indicates an acceptable level of performance and cost for the flywheel, the PNGV should plan for the physical installation of flywheel hardware in a post-2000 concept vehicle.

POWER ELECTRONICS AND ELECTRICAL SYSTEMS

All three USCAR partners have elected to pursue HEV designs for the 2000 concept vehicle (Malcolm, 1997). Fuel cells for energy generation and flywheels for energy storage could very well be practical within the PNGV time frame. Electrification of major auxiliary functions, e.g., air conditioning and power steering, is attractive for ease of control and improved energy management and efficiency. Success depends on the development of efficient and economically acceptable actuators, motors, and power electronic converters.

Program Status and Progress

The electrical and electronics power conversion devices team (EE technical team) has made considerable progress in organizing and coordinating its efforts. The committee had noted in the third review that this team lacked leadership and had recommended that a full-time leader be appointed (NRC, 1997). Not only has this been done, but two technical subteams have also been established to address power electronics enablers and electric motor enablers. The subteams appear to be effectively setting priorities and addressing the important issues in their areas. Both have made designing and manufacturing for low cost their highest priority.

The progress of the EE technical team since the committee's last review is evident in the team's performance as measured against the targets for specific power, volumetric power density, cost, and efficiency. Except for motor efficiency, the 1997 targets for both the motor and electronics are reported to have been either met or exceeded (Malcolm, 1997). The reported cost of $15/kW for the power electronics module is particularly impressive in comparison to the 1997 target of $25/kW and is extraordinary in comparison to the interim technical target of $100/kW by 1997 that was indicated in the PNGV Technical Roadmap (PNGV, 1997).

The EE technical team has also made progress in leveraging the activities of other organizations, especially the Office of Naval Research (ONR), Wright-Patterson Air Force Base, and the National Renewable Energy Laboratory. The success of the PNGV power electronics developments appears to rely heavily on ONR's Power Electronic Building Block (PEBB) program. The EE technical team has established a liaison with the PEBB researchers and has been working with the PEBB program management to focus attention on PEBB specifications that are applicable to the PNGV program.

Although the EE team has made considerable progress in working more closely with the systems analysis team, some necessary models have still not been provided to the systems analysis team, including models for motor/generators, power electronic converters, and control algorithms for both the series and parallel hybrid drive configurations. The EE technical team is currently working on providing them.

Accessory loads, heating, ventilation, and air conditioning (HVAC), and regenerative braking were identified by the EE technical team as priorities for technology development. Work is being done on starting/charging and accessory loads by both the USCAR partners and their suppliers. In the committee's third review, regenerative braking efficiency was identified as an area of concern (NRC, 1997). Because this topic was not addressed in presentations for this (fourth) review, the committee has concluded that regenerative braking efficiency is still an outstanding issue. The EE technical team is aware of HVAC-related work being done at the national laboratories and is coordinating the electrical requirements of the HVAC system with developments as they are evaluated and adopted by the systems analysis team and the systems engineering team.

The USCAR partners have independently chosen HEV designs for the year 2000 concept vehicle. This means that each company may have different electrical system architectures and requirements.

Assessment of the Program

Both motor and power electronics technology will meet PNGV functional requirements. The remaining challenge for the EE technical team, as the team has clearly stated, is meeting the cost targets for these components. The 2004 target cost for motors is $4/kW, which the team recognizes as the approximate cost of materials in state-of-the-art motors. Although the team apparently already has achieved the very low cost of $15/kW for power electronics, the PNGV Technical Roadmap target for 2004 requires a further reduction of more than 50 percent to $7/kW. In the committee's opinion, the impact of the cost of power electronics on other vehicle systems has either not been recognized or has been considerably underestimated. Although the EE technical team has tried to provide models to the systems analysis team, coordination with other program teams could still be improved.

The committee was not made aware of assessments of foreign technology by the EE technical team. However, given the introduction of commercial HEVs by foreign manufacturers (e.g., the Toyota Prius) in the past year, the committee believes that the USCAR partners are aware of these developments and have been conducting independent evaluations of foreign technology.

The hybrid designs chosen independently by the USCAR partners will make a more comprehensive evaluation of competing technologies possible in an application where no single correct approach is obvious, especially for the electrical and electronic subsystems. However, the committee has not seen any evidence that the cost target of $7/kW for electronic power modules can be achieved in any of the designs. Given that the cost of less sophisticated computer power supplies produced in high volume is about $100/kW, the committee is concerned that the cost goals established by the EE technical team are too aggressive. The committee is also concerned that the EE technical team's reliance on the ONR's PEBB program will jeopardize the cost targets if the PEBB program changes direction, loses funding, or cannot meet its goals.

Recommendations

Recommendation. The PNGV should perform a thorough and convincing verification of the reported 1997 cost of power electronics in conjunction with a reevaluation of the 2004 cost goal. The reevaluation should include identifying developments that support the assumption that the cost target can be met.

Recommendation. The PNGV electronics and electrical systems team's reliance on the Navy's Power Electronic Building Block (PEBB) program should be mitigated by the initiation of a PNGV-specific cost reduction program.

Recommendation. Given the critical importance of electrical and electronic systems to the success of the PNGV program, the electrical and electronic systems technical team should provide cost models to the systems analysis team as soon as possible to ensure that cost assumptions for other subsystems that rely on power electronics are consistent with the projections and targets.

MATERIALS

The reduction of vehicle mass through the use of lightweight materials is one of the key elements in meeting the fuel economy goal of the PNGV program. Because the leading candidate materials are currently more costly than the steel used today, the requirement that there should be no increase in the overall cost of vehicle ownership presents a major hurdle to meeting the fuel economy goal.

PNGV personnel are working closely with materials suppliers to develop less costly manufacturing processes and new design practices that utilize materials more efficiently. Each lightweight material alternative also offers a different weight reduction, has different costs, and raises different issues in terms of manufacturing feasibility, design experience and confidence, infrastructure needs, new failure modes, repairability, and recyclability.

To meet the 80 mpg vehicle fuel economy objective of Goal 3 while maintaining vehicle performance, size, utility, and cost of ownership, vehicle curb weight will have to be reduced by 40 percent from 3,240 to 1,960 lb. Table 2-3 shows the breakdown of the weight reduction targets by major vehicle subsystem.

Program Status

The major alternatives under consideration in 1997 for reducing vehicle weight were: more efficient design of the current steel-intensive vehicle, which is being led by the American Iron and Steel Institute; aluminum sheet and castings; fiber-reinforced composites; magnesium; matrix composites; titanium; and lightweight glazing (thinner glass and polymers).

More Efficient Steel Design

The approach of the American Iron and Steel Institute program has the potential of a 20 percent weight reduction for the body-in-white (BIW) structure (bolt-on panels, such as the hood, doors, front fenders, and deck lids are not included in the BIW). Based on efficient steel design technology, the weight of the baseline PNGV vehicle BIW could be reduced from 598 to 478 lbs; the cost could be reduced by $154. The PNGV material team also studied the possibility of using a stainless steel space frame but found that it would generate a cost penalty of $200 for a weight saving of only 22 percent, which is half of the savings needed for the Goal 3 vehicle.

Aluminum and Magnesium

The total potential vehicle weight savings with aluminum sheets and castings is 600 lbs; the total potential weight savings with magnesium castings alone is 150 lbs. Note that the weight saving potentials for the alternative materials are not additive because certain parts, wheels for example, have been targeted for both materials in the individual computations. On a part-by-part basis, the weight saved by substituting steel sheet with aluminum sheet is typically more than 50 percent. This number does not include secondary weight savings that result from reducing the size of other components, such as brakes, wheels, and suspensions. The total vehicle must be redesigned around the primary weight saving to assess accurately potential secondary savings.

TABLE 2-3   Vehicle Weight Reduction Targets for the Goal 3 Vehicle

Subsystem	Current Vehicle (lbs)	PNGV Vehicle Target (lbs)	% Mass Reduction
Body	1,134	566	50
Chassis	1,101	550	50
Power train	868	781	10
Fuel/other	137	63	55
Curb Weight	3240	1960	40

Source: Stuef (1997).

The automotive industry has considerable production experience with aluminum in the form of stamped body panels. The procedures and processes for recycling aluminum are also in place today, and much of the material is returned to high value automotive applications. As a matter of public record, several aluminum-intensive prototype vehicles have been built outside the PNGV program by the USCAR partners and evaluated for ride, handling, NVH, crashworthiness, joining, and painting (Jewett, 1997). Thus, the change to an aluminum-intensive vehicle would not be a major technology challenge because the USCAR partners already have extensive design and manufacturing expertise with this material. The challenge is to develop new processing methods so that an aluminum-intensive vehicle can be made as inexpensively as a steel vehicle. Based on a price for aluminum of $1.60/lb, the cost penalty of an aluminum BIW is estimated at $400.

Cast aluminum and magnesium would be used in the chassis, the body, and the power train subsystems. Major efforts to reduce the costs of feedstock and improve the casting processes of both materials are under way. Studies to improve the machinability of magnesium castings and to develop a lower cost high-temperature alloy are also under way. Improved processes for recycling magnesium will have to be developed.

Fiber-Reinforced Plastic

In 1997, a number of serious obstacles were identified that will limit the use of graphite fiber-reinforced plastic composite material (GrFRP) in meeting Goal 3. First, not enough is known about the consistency of the mechanical properties of GrFRP when produced in large volumes. Second, the criteria and methods for reliably designing GrFRPs for fatigue resistance and crashworthiness in the complex loading environment presented by the automotive body structure are far from production-ready. Third, methods for recycling the material into high-value applications to take advantage of its intrinsic properties, as opposed to using it only as filler, must still be developed.

According to the latest PNGV/USCAR studies, the potential BIW weight reductions with GrFRPs are only a few percentage points better than for aluminum, 59 percent versus 55 percent. The committee was surprised at the reportedly small margin of improvement of GrFRPs over aluminum. PNGV/USCAR explained that the small margin reflected a combination of two factors. Manufacturing considerations and the need for sufficient structural strength to accommodate multi-axial loads required more material than the committee had expected.

The major barriers to the intensive use of GrFRPs for a Goal 3 vehicle, however, are the high cost of the graphite fibers and the lack of a suitable high-volume manufacturing process for the material. Currently, the cost penalty of a GrFRP BIW is higher than a steel BIW, and there is no feasible high-volume manufacturing process for the thin sections.

Glass fiber-reinforced plastic composites (FRPs) offer weight savings in the 25 to 35 percent range. FRPs that incorporate thermoset resins have been used extensively in noncritical stressed structures, such as hoods, deck lids, doors, and fenders. One drawback of using thermoset materials is the substantial investment in tooling because of the relatively slow cycle times. Chrysler is investigating low-cycle-time injection molding for glass-reinforced thermoplastic resins, which has additives for improving crashworthiness and weather resistance. Chrysler is considering using them in both body panels and body structures. Computer simulation and hardware tests are being used to test the crashworthiness of these materials.

Metal Matrix Composites

The applications for metal matrix composites are in the chassis and power train subsystems. The total potential weight savings of using metal matrix composites is only 30 to 50 lbs. The major hurdles to developing applications of this material are feedstock costs and the development of a reliable process for compositing the materials.

Titanium

The applications for titanium are in the chassis (40 lbs potential savings) and power train (10 lbs potential savings). The components of interest are springs, piston pins, connecting rods, and valves. The major barrier to the use of this material is its high feedstock cost.

Lightweight Glazing

New glazing materials—thin glass and polymers—are being considered for weight reduction. The potential weight saving is 50 lbs. The major concerns for polymers are abrasion/scratch resistance and cost.

Program Progress and Plans

Steel

The American Iron and Steel Institute efficient steel vehicle design program, which will be completed in 1998, offers near-term weight and cost savings that can be implemented in goal 1 and 2 applications, once the USCAR partners have verified the findings. However, the intensive use of steel is not feasible for meeting the Goal 3 weight reduction targets unless there is a major breakthrough in power train efficiency.

Aluminum Sheet

In the past year, aluminum has become a leading structural material candidate for the Goal 3 vehicle technology selection process because (1) it offers a much larger percentage weight reduction than steel, (2) it is much less costly than GrFRP, (3) the knowledge base for the design and manufacture of this material is extensive, and (4) existing stamping facilities for steel can be used for aluminum without major modifications.

Because of aluminum's importance to weight reduction and high fuel economy, a major cooperative research and development agreement (CRADA) has been initiated between Reynolds Metals, LANL, and the USCAR United States Automotive Materials Partnership to reduce the cost of aluminum sheet through the development of a thin-slab (less than 1 in thick) continuous-casting process to replace the more costly ingot-based process used today. The preliminary results of this program are very promising. Another program to develop low-cost, non-heat-treatable alloys competitive with the 6000 series aluminum alloys is also under way, as well as studies on improving the formability of aluminum stampings and making the walls of aluminum extrusions thinner. The goal of all of these programs is to reduce the cost of an aluminum-intensive vehicle.

Graphite-Fiber-Reinforced Plastic Composite Material

The weight saving of GrFRPs compared to aluminum is not sufficiently attractive for this material to be the leading candidate in the Goal 3 technology selection process. Cost and thin-section manufacturing are major issues that cannot be resolved in time to support Goal 3 year 2000 concept vehicles. The development of GrFRPs should be continued with applications targeted beyond Goal 3. R&D should concentrate on reducing the fiber cost to $3/lb, developing a process for mass producing components in thin sections, acquiring a deeper understanding of reliable design for complex loading conditions, and developing high-value applications for the recycled material.

The injection molding of FRPs (glass-reinforced thermoplastics) being investigated by Chrysler looks much more promising at this juncture. Although the potential weight saving of 25 to 35 percent is not as great as it would be with GrFRPs, FRP may compete effectively with aluminum for body panels and certain structural applications because the processing technology could integrate two or more parts that are currently made with steel or aluminum into a single part. Nevertheless, the extensive design and production experience with aluminum for body panels and structural applications gives FRPs an edge in the near term.

Aluminum and Magnesium Castings

Several important R&D studies were conducted in 1997 to reduce the costs and improve the properties of aluminum and magnesium castings, including sand casting, semi-permanent mold casting, squeeze casting, and high-pressure die casting. Simulation models of the casting flow and solidification processes are being developed to predict microporosity as a function of the part and casting process design. Improved nondestructive evaluation techniques are being developed to support the use of aluminum and magnesium in more demanding applications. The development of rapid tooling processes for die-casting applications is under way. Methods and materials for improving the die-casting dies are also being studied to reduce the overall cost of using these materials.

Plans for 1998

The plans for 1998 are to continue the major material initiatives already under way. The committee agrees that the focus should be on following through on the cost-reduction initiatives begun in 1997, especially for the continuous casting of aluminum sheet.

Assessment of the Program

In 1997, the PNGV materials technology team and the vehicle engineering team made a thorough joint evaluation of the lightweight material candidates for the Goal 3 technology selection process. The criteria included potential weight savings, feedstock cost, manufacturing cost and feasibility, design and manufacturing experience, and the ability to recycle the material into high-value applications. The committee agrees with the criteria for selection used by the PNGV teams and with their conclusion that aluminum is the lightweight material of choice for intensive use in support of Goal 3 objectives, along with the selective use of FRPs, magnesium, GrFRPs, and titanium. The committee also agrees with PNGV/USCAR's assessment that intensive use of GrFRPs should be a longer range goal.

The manufacture of an aluminum-intensive vehicle would not meet the Goal 3 cost objectives if the aluminum were produced with today's mill practice technology. Consequently, the CRADA to develop continuous casting of aluminum sheet for automotive applications is critical to eliminating the cost penalty and, therefore, critical to meeting the overall objectives of Goal 3. Finding ways to improve the physical properties and reduce the cost of cast aluminum and magnesium is also important to meeting Goal 3 objectives. PNGV/USCAR should monitor these projects closely and ensure that they have adequate resources to meet these critical objectives.

Recommendations

Recommendation. The development of the continuous casting process of aluminum sheet should be given the highest priority in terms of resources and technical support. This includes support of the work already under way to characterize the material in terms of its microstructure, strength, ductility, formability, and weldability in parallel with the development of mill processing techniques.

Recommendation. The development program for low-cost graphite fiber should be continued for longer-term applications beyond Goal 3. A new program for manufacturing graphite-fiber-reinforced plastic composite materials in thin sections should be initiated to take advantage of the unique properties of this material.

3

Systems Analysis

The PNGV Technical Roadmap, which is guiding the technical teams, states the following in Section III-A (PNGV, 1997):

> The role of systems analysis in the PNGV is to support component, systems, and vehicle development by providing the analytical capability to efficiently and accurately assess competing technologies, and vehicle concepts against Goal 3 objectives and vehicle performance requirements. This will enable an objective evaluation of cost, benefit, and risk, in order to focus on the best options, with the least expenditure of resources.

A vehicle system model has been created by the systems analysis team with the assistance of TASC/Southwest Research Institute (SWRI) consultants. The model makes it possible to compare the relative performances of selected vehicle configurations. A good example of its utility is in comparing the performance of the HEV configurations. The HEV has two significant advantages that contribute to fuel economy: the potential to recover some braking energy; and the ability to run the selected power plant in a restricted, more efficient load and speed range. Disadvantages of the HEV include complexity, weight, and cost. The system models provide a means for assessing objectively the relative performance of each HEV configuration.

Systems analysis, based on effective computer modeling tools, is the most efficient way to ensure the optimization of vehicle performance for selected vehicle configurations. It also provides the opportunity to study trade-offs between candidate subsystems during the technology selection process. The models facilitate the preparation of specifications for each candidate subsystem and support the establishment of engineering targets. The ongoing technology selection process for demonstration vehicles will depend heavily on the results of systems analyses.

All but one of the committee's recommendations in the third report were addressed by the PNGV systems analysis technical team. The exception was that an assessment be made of the value of the expenditures during 1996 for the TASC/SWRI contractor team. During the committee's present review, the team outlined its accomplishments and its plans for 1998. The committee concluded that considerable progress has been made in the past year and that effective systems analysis is now in place to provide design support to the technical teams. The vehicle and subsystem models being created should effectively support the technology selection process. Table 3-1 outlines the systems analysis team's responses to the recommendations and observations in the committee's third report. In general, the responses, combined with presentations to the committee, indicated that the team had responded effectively to the committee's concerns. The major exception was that cost and reliability models have not been reviewed with the committee. The committee believes that both models are necessary to the effective selection of technologies for concept vehicles that will meet the program's objectives.

PROGRAM STATUS AND FUTURE PLANS

Modeling

During the committee's review, the systems analysis technical team made a presentation describing the computer models being developed, showing typical simulation assumptions for specific vehicle configurations, detailing data on fuel economy for parallel and series hybrid vehicles, and describing the sensitivity of fuel economy to changes in vehicle mass. Detailed charts and supporting analysis were presented summarizing the timing and status of subsystem models for all the subsystems being considered. As expected, the model results varied widely, from the internal combustion engine, for which model predictions correlate well with actual data, to the Stirling engine, for which a good model may not yet exist. The PNGV management stated that the precision of the systems analysis tools, in general, exceeds the accuracy of data available for each subsystem. The committee understood this to mean that the analytical models are very powerful and complete and that the simulation models are accurate, provided that they are based on adequate subsystem and component performance data. The PNGV must now build and test subsystems and vehicles that will provide performance data to compare to the performance predicted in the models.

A summary of technical tasks and timing was also presented, and the committee was given detailed plans and schedules for each model requirement indicating the status of each subsystem relative to selected key parameters, including performance, efficiency, emissions, thermal and load transients, heat rejection, scaling, volume and aspect ratio, ride and handling, cost, and reliability. These

TABLE 3-1 Response of the PNGV Systems Analysis Team to the Committee's Third Report

Observations and/or Recommendations	Response
1. The pace of systems analysis must be accelerated and funding resources identified.	USCAR and the U.S. Department of Energy have increased resources and established a long-term funding plan.
2. A lack of good model validation data is hindering systems analysis.	Resources to obtain both vehicle and component data from subsystem and vehicle development sources have been committed.
3. Cost and reliability models are weak.	The next version of the computer model includes a refined cost model. The reliability model is adequate, and model validation now depends on good experimental data.
4. More effort should be focused on the effective use of models by the technical teams.	The systems analysis team is concentrating on designing and developing an effective graphical user interface to meet the needs of all the technical teams.
5. Greater participation and interaction with the engineering teams are necessary.	The systems analysis steering group participates in the meetings of the PNGV engineering team and addresses technical concerns.
6. Systems analysis should focus on technologies that have a high probability of success in the year 2004.	Model refinements in the next phase of the program will focus on the most promising technologies, as defined by the technology selection process.
7. Detailed systems studies of component technologies are needed to define system and subsystem requirements from a vehicle perspective.	Analytical studies respond to the requirements established during the meetings with the technical teams. Models are based on these analyses.
8. Systems studies are needed to confirm the accuracy of the component objectives detailed in the PNGV Technical Roadmap.	The PNGV Technical Roadmap is continually updated in response to systems analysis and modeling. A major update will be made after the technology selection process is completed in 1997.
9. Competitive technology assessments, especially of foreign technology, should be routinely conducted.	All of the automotive companies involved routinely monitor competitive activity to ensure their competitiveness.

plans described the interactions of the system analysis team with the other technical teams in modeling and supporting design analyses. The committee had been very critical in the third review about the lack of interaction and was gratified to see that the situation is much improved (NRC, 1997). This was confirmed by remarks made during review presentations by other technical teams. The committee is still very concerned about the status of subsystems like the fuel cell, the battery, and the power electronics module, for which reliable models are still evolving. The committee urges the systems analysis team to increase its efforts to create models for these subsystems.

The technology selection process will depend largely on the computer models, and the committee's review of the preparation for technology selection revealed that models are being used effectively. The most promising models have been identified and vehicle level control strategies are being developed.

Future Plans

The following PNGV development needs beyond November 1997 were provided to the committee during its review (Kenny et al., 1997).

(1) Nonperformance model enhancements
    refine cost model
    validate reliability model
    integrate optimization capability
(2) Power bus formulation
    update component models (e.g., switch reluctance motor)
    validate data for component models
    improve parallel control algorithms (e.g., user definable)
    improve series control algorithms
(3) Model architecture
    limit client server (1 PC/1 workstation)
(4) Voltage bus formulation
    create motor and generator equivalent circuit models
    develop power electronics model
    develop new control algorithms
    update component models (focus on transient response)
(5) Additional capabilities not being collaboratively pursued
    manufacturability
    recyclability
    packaging/crashworthiness
    noise, vibration, harshness (NVH)
(6) Integration of modeling of subsystems (fuel cells, batteries, power electronics modules) with modeling of the entire vehicle system

AREAS OF CONCERN

During the committee's review, it was obvious that very little cost modeling had been done by the PNGV systems analysis team. The committee was informed that provisions for cost modeling had been made, but little has been done so far. The committee believes that probabilistic models of vehicle and subsystem costs, with confidence levels, will be very important for the technology selection process. The computer model should provide analysis and presentation of a complete cost picture for every vehicle configuration selected. Cost is still a major barrier to satisfying the original PNGV affordability objective.

The PNGV is fully aware of the challenge it faces in achieving the Goal 3 cost objectives, particularly for hybrid vehicles. The committee raised this question with the PNGV, and two of the USCAR partners volunteered to share proprietary cost data with a subgroup of the committee. The data reviewed satisfied the committee that good cost estimates are available for subsystems like the CIDI hybrid vehicle, for which design and manufacturing technology has been developed. Cost estimates for mass producing more advanced subsystems, like fuel cells, are not adequately developed for accurate cost forecasting.

Nevertheless, the committee believes it is imperative that the PNGV create complete cost models for all of the subsystems under consideration so that effective cost trade-offs can be considered. Even if the models are not as accurate as desired, they will at least provide baselines for comparing all of the systems under consideration and will highlight the specific developments necessary to realizing acceptable costs.

Without compromising the proprietary interests of the USCAR partners, the results of their analyses should be communicated to the systems analysis team for use in their cost models. Furthermore, the USCAR partners should agree on representative subsystem target costs, which are important to guide the development efforts of the many participants in the PNGV program.

Another area that deserves more attention is reliability models. Vehicle design life and safety are major considerations in meeting vehicle requirements. Each subsystem being considered introduces new failure modes that could affect design life and safety. The systems analysis team should be deliberately structuring component and subsystem reliability models that can be used to predict vehicle reliability during the technology selection process and then as a basis for hardware test verification. The PNGV's reliability model(s) should be of individual components, subsystems, and the total vehicle operation in a defined environment. The models should be statistical and should define failure modes and effects. Each component should have a design reliability model based on specific assumptions and test data.

RECOMMENDATIONS

Recommendation. Systems analysis and computer modeling are essential tools for making system trade-offs and optimizing performance. The PNGV should create detailed, rigorous cost and design reliability models as soon as possible to support ongoing technology selection. These models should be continually upgraded as new information becomes available.

Recommendation. Because cost is a significant challenge to the PNGV, the USCAR partners should continue to conduct in-depth cost analyses and to use the results to guide new development initiatives on components and subsystems.

4

Technology Selection for Concept Vehicles

Goal 3 in the PNGV program plan calls for the development of concept vehicles from 1997 to 2000 (PNGV, 1995). A concept vehicle is one that, at the outset, is projected to be capable of meeting most of the critical vehicle attribute parameters set forth in the stated goal. It is a "proof of concept" vehicle that demonstrates technical feasibility. A concept vehicle may, however, incorporate components and manufacturing techniques that are not suitable for mass production and may cost much more than would be permissible in a commercially viable product. If so, however, a credible plan should have been made for reducing the cost of all critical components so that the projected overall cost meets the ultimate goal. Table 4-1 lists the Goal 3 parameters specified by the PNGV for a vehicle with up to triple the fuel efficiency of the baseline vehicles. Between 2000 and 2004, the program plan calls for the development of production prototype vehicles that, in addition to meeting all of the stated parameters, could be mass produced at a competitive cost.

Making technology selections based only upon meeting Goal 3 *as stated* may not produce the anticipated societal benefits. First, Goal 3 specifies a target fuel efficiency only for a certain class of vehicles—high-volume, midsize passenger cars. The target is not to reduce overall, nationwide passenger car energy use or emissions. For the societal benefits of increased fuel efficiency to be fully realized, other segments of the passenger car fleet, such as light trucks, must also be affected by the developments under the PNGV program. Second, Goal 3 does not take into account that major technological changes in mass produced cars are very likely to have significant secondary effects on energy use and emissions. The production and operation of passenger cars and light trucks accounts for a sizable portion of the raw materials used by all industries. Radical changes in

TABLE 4-1   Attributes of the PNGV Goal 3 Vehicle

Vehicle Attributes	Parameters
Acceleration	0 to 60 mph in 12 seconds
Number of passengers	up to 6
Operating life	100,000 miles (minimum)
Range	380 miles on 1994 combined drive cycle
Emissions	meet or exceed EPA Tier II requirements
Luggage capacity	16.8 ft^3, 200 lbs
Recyclability	80 percent
Safety	meet federal motor vehicle safety standards (FMVSS)
Utility, comfort, ride, handling	equivalent to current vehicles
Purchase and operating cost	equivalent to current vehicles when adjusted for economics

Note: Utility refers to the degree to which a given vehicle is useful to an individual car buyer and includes attributes such as passenger space, trunk capacity, seating capacity, and ergonomics.

these raw materials will have secondary effects on emissions from stationary sources, energy usage, and other parameters in a wide variety of industrial operations. To determine the overall societal effects, the direct and indirect effects of these changes on the overall industrial system will have to be analyzed.

METHODOLOGY AND RESULTS OF TECHNOLOGY SELECTION

A large number of technologies must be considered to reach a goal as ambitious as Goal 3. During the first four years of the PNGV program, both the USCAR partners and the government research managers examined hundreds of technologies and ideas that might have contributed to the success of the program. The 2004 deadline for completing production-ready prototypes, together with reasonable limits on available resources, dictated that selections of the most promising technologies be made during 1997. As the PNGV program has matured, however, it has become evident that some promising technologies will not be ready for the construction of a production prototype car in 2004. This suggests that the technology selection process should recognize some technical approaches as near-term candidates, some as longer-term possibilities, and some as not likely to be used in passenger cars in the foreseeable future, although they may be deserving of continued R&D at some level.

The PNGV reached its initial technology selection process milestone on schedule, and the USCAR partners can now continue with the design and construction of concept vehicles. Meeting the PNGV goals required that they adhere to this demanding schedule, and they have accomplished this task. The committee notes and commends their progress. The vehicle criteria in Table 4-2 were the basis for deciding which technologies will be applicable for concept vehicles.

Although the technology selection has been completed, the process PNGV

TABLE 4-2   PNGV Criteria for Year 2000 Concept Vehicle
Technology Selection

Category	Criteria
Safety	must meet FMVSS standards by design
Fuel economy	up to 80 mpg
Emissions	capable of meeting Tier II gaseous and applicable ultra-low emission vehicle (ULEV) particulate standards by 2004
Performance	within ± 30 percent of Goal 3 targets
Utility	within ± 30 percent of Goal 3 targets
Cost potential	within 30 percent of baseline vehicle

used and the way the criteria were established are not entirely clear to the committee. Given the fact that the fuel economy target is "up to" three times baseline fuel efficiency and the possibility that all of the Goal 3 criteria might not be met simultaneously, several basic selection criteria, either alone or in combination, could have been used to guide the PNGV's decisions. Possible ways to choose technologies are listed below:

- Select technologies that, with a reasonable stretch, are very likely to approach or meet the 80 mpg goal.
- Select technologies that are expected to motivate major technological advances.
- Select technologies that will allow the testing of many individual system components in an overall system.
- Select technologies that may allow the U.S. industry to leapfrog its international competitors.
- Select technologies that are likely to result in marketable vehicles.
- Select technologies that could lead to vehicles with a large aggregate impact on energy consumption, emissions, global warming, or other environmental concerns.

Some technologies might be chosen under all of these selection concepts, but others may be consistent with only some of them. Even if the same technologies would have been chosen at this stage, a statement explaining how the selection criteria were established and used in relation to the requirements of Goal 3 would be helpful for allocating resources among the technologies that require more R&D. As concept cars are proposed, they will inevitably emphasize some selection criteria more than others.

The USCAR partners made extensive use of a simulation model in deciding which technologies to emphasize in the next phases of the program. Based on control algorithms, power train power output, and fuel consumption data, together with vehicle and chassis parameters, this simulation predicts the fuel economy and acceleration performance of a hypothetical vehicle. The simulation has been validated by tests with conventional power trains and found to be acceptable. The accuracy of predictions for new configurations is limited primarily by how well the characteristics of the new components can be modeled and their behavior forecast. For technology in an early stage of development, such as fuel-cell power plants, subsystem specifications are not detailed enough to provide accurate performance predictions.

For the first-pass technology selection, simulated comparisons of candidate power train technologies did not account for potential mass differences of either the power train or the chassis in response to variations in subsystem mass. Although achieving the vehicle acceleration and fuel economy targets is essential, these measures alone are not sufficient for choosing a power train. Exhaust emissions, fuel storage, and vehicle refueling are also critical. A satisfactory analytic technique for making detailed comparisons of emissions by the power plants currently being considered has not been developed. Therefore, at this stage, comparisons can only be based on engineering judgments.

Vehicle components must be packaged in a way that will accommodate the overall allowable vehicle dimensions and at the same time provide sufficient space for passengers and luggage. This criterion can only be met after component dimensions and mounting requirements have been well established. Packaging studies have been completed for some, but not all, power train configurations.

Projecting the cost of experimental systems is extremely difficult, particularly when completely new components and technologies are involved. For some components, realistic costs cannot be projected because high-volume manufacturing processes have not been developed or even conceived. This is probably the weakest link in the judgment chain for technology selection.

Finally, making trade-off decisions about the technologies that should receive further attention at this stage requires subjective judgments about the likelihood of future breakthroughs and the ultimate marketability of the resulting vehicles. Marketability, in addition to the obvious criteria of overall selling price, performance, durability, reliability, and vehicle appearance, also depends on several less obvious factors, such as insurability, recyclability, and excessive product liability exposure.

Table 4-3 is a list of the most promising technologies selected by the PNGV (see Chapter 2 for more details on individual technologies), grouped by type of technology. Some of these technologies will be used in early concept vehicles for the year 2000; others will continue to be developed for use in post-2000 concept vehicles. No priorities have yet been assigned to these technologies based on the likelihood of their meeting PNGV Goal 3 targets.

A major consideration in PNGV's technology selection for the 2000 concept vehicle was the selection of the energy converter and its related fuel efficiency and exhaust emissions. Through computer simulation, PNGV has computed the expected fuel efficiency of a lightweight, high-efficiency vehicle for various

TABLE 4-3   Most Promising Technologies Selected by the PNGV in 1997

Category	Technical Area and/or Technology
Power trains	parallel hybrid electric drive
Energy converters	CIDI engine fuel cells
Energy storage	nickel metal hydride batteries lithium batteries
Emission controls	lean NO_x catalyst exhaust gas recirculation particulate traps
Fuels	fuel with less than 50 ppm sulfur Fischer-Tropsch fuel dimethyl ether fuel
Electrical systems and electronics	induction, reluctance, permanent-magnet motors PEBB, IGBT, MOSFET, MCT semiconductors ultracapacitors
Materials	aluminum and/or reinforced composite body-in-white
Reducing energy loss	low rolling resistance tires reduced HVAC requirements and more efficient components

Source: Based on York (1997) and the PNGV Technology Selection Announcement (see Appendix F).

Note: PEBB = power electronic building block. HVAC = heating, ventilation and air conditioning. IGBT = insulated gate bipolar transistor. MOSFET = metal oxide semiconductor field effect transistor. MCT = MOS (metal oxide semiconductor) controlled thyristor. Body-in-white constitutes the primary structural frame of the vehicle, not including bolt-on pieces, such as the hood, doors, front fenders, and deck lids.

energy converters and power train combinations (see Figure 4-1) and compared them to conventional power trains. The respective fuel economies ranged from 27 mpg to more than 80 mpg. Uncertainties in the estimates for each power train are represented by the length of the horizontal bars in Figure 4-1.

The selection of the concept vehicle energy converter was based on a number of factors, including overall performance, fuel efficiency, emissions, cost, size, weight and state of development. Each company is giving near-term priority to the 4SDI internal combustion engine. Figure 4-2 shows the four types of internal combustion engines that were analyzed. The stratified-charge CIDI engine was ultimately chosen for its potential to achieve a thermal efficiency 23 to 28 percentage points greater than a baseline engine. This was the highest efficiency of all of the 4SDI engines. Nevertheless, the CIDI engine still faces major challenges in meeting the NO_x and particulate emissions targets (see Chapter 2). The advanced gas turbine and the Stirling engines were not selected because they either have lower levels of performance, are less mature technologies, or have higher projected production costs.

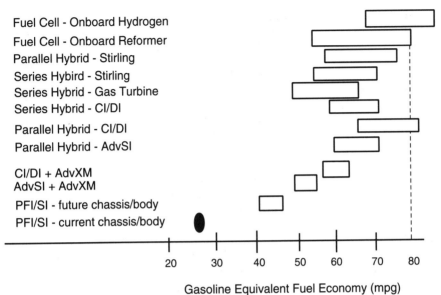

FIGURE 4-1   Relative fuel economy projections for various vehicle/power train configurations. All configurations except the current chassis/body include PNGV-class lightweight body, chassis, and interior (2,000 lbs); advanced aerodynamics; and low rolling resistance tires. The PFI/SI (port fuel injection/spark ignition) conventional vehicle is the baseline. Key: ● = current vehicle chassis/body; ☐ = future efficient vehicle body/chassis [light weight (2,000 lbs), sleek aerodynamics and low rolling resistance]. Variance denotes downward uncertainty from unmodeled energy losses and upward uncertainty from improvement as technology matures. CIDI = compression ignition/direct injection; AdvSI = advanced spark ignition. AdvXM = advanced transmission. Source: Provided to the committee by PNGV.

These analyses (Figures 4-1 and 4-2) indicate that an HEV with a CIDI engine is the most appropriate near-term (year 2000) choice, although the fuel-cell energy converter is clearly the longer-term choice, provided the many challenges facing it can be overcome. The committee agrees with the PNGV's technology selections based on the performance potential and state of development of the systems and subsystems.

CONCEPT VEHICLES

The USCAR partners have independently built test vehicles that incorporate some of the advanced technologies and components being considered and have shared the results of their evaluations. In the future, however, each company plans to build its own concept vehicle or vehicles. The overall design and construction

68

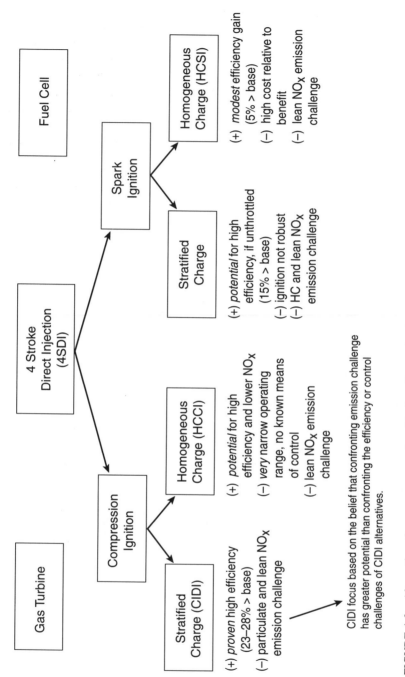

FIGURE 4-2 Alternatives for energy conversion. The baseline engine is the port fuel injection homogeneous charge spark injection for model year 1993. Source: 4SDI Technical Team (1997).

of these vehicles may go well beyond the precompetitive R&D boundaries that define USCAR's cooperative work. Because each manufacturer views the market somewhat differently, each is expected to develop a different type of concept vehicle. They will also have different approaches to mass reduction and will use different power train configurations. All three manufacturers have programs to develop hybrid-electric power trains, and many different configurations and degrees of hybridization are possible. Also, in the next few years, some test-bed cars that incorporate one or another advanced concept will continue to be built and tested, in addition to vehicles that meet all of the PNGV concept vehicle criteria.

In spite of this diversity, many areas of precompetitive research have still not been explored. All of the most promising technologies listed above still require major development before they will be serious candidates for use in mass-produced passenger cars and trucks. Reducing the cost of these technologies is the most common concern.

POST-2000 CONCEPT VEHICLES

Some of the technologies being explored (e.g., fuel-cell power plant, advanced batteries, ultracapacitors, and flywheels) are very promising but are unlikely to be ready for application in full-concept vehicles before 2000. These technologies are expected to be developed under the PNGV cooperative program and incorporated in post-2000 concept vehicles.

HYBRID VEHICLE DEVELOPMENT

All three USCAR partners have DOE-supported HEV programs that started at different times and have somewhat different goals. The vehicles being constructed under these contracts should not be confused with the concept vehicles being built for the PNGV program. They will not necessarily meet the criteria for the PNGV concept vehicles. For example, they were conceived as vehicles that would deliver only twice the baseline fuel economy. Nevertheless, these programs will speed the development of components specifically suited for HEV applications and increase the expertise in systems management issues associated with this complex power train.

RECOMMENDATION

Recommendation. The relationship between the criteria for technology selection and the critical requirements of Goal 3 should be made more explicit to facilitate the proper distribution of resources for an ongoing, well structured research and development program.

5

Fuel Strategy

Throughout the development of the automobile, engine and fuel advancements have been closely linked. The performance and emissions of the sophisticated power plants in current automobiles are critically dependent on the widespread availability of fuels that are tailored specifically for them. Automobile manufacturers and fuel suppliers have a common bond that is crucial to their business successæthey must both please the same customer. Historically, this marketplace interdependence and performance linkage (e.g., the phaseout of lead in gasoline) have meant that engine and fuel changes had to be carefully coordinated to maintain customer satisfaction.

The PNGV Goal 3 does not deal explicitly with secondary interactions, and the program initially focused exclusively on the vehicle. Although this shortcoming was noted by the committee in previous reports, no consensus by the PNGV members has been reached on weighing trade-offs between the energy consumed or emissions produced by the vehicle and the energy consumed or emissions occasioned by fuel processing and distribution. However, the critical role of the transportation fuel supply infrastructure has been acknowledged by PNGV now that fuel cells and CIDI engines have emerged as primary power-plant concepts (see Appendix F). Recent concerns about reducing fine particulate emissions from CIDI engines have brought the fuel issue into sharp focus. (The relationship between CIDI engine development and fuel developments is discussed in the section on Internal Combustion Reciprocating Engines in Chapter 2.)

POTENTIAL FUEL MODIFICATIONS AND BENEFITS

Many examples could be cited showing the close links between automotive engines and fuels. The performance and durability of spark-ignition engines are critically dependent on fuel octane rating; diesel engines are similarly dependent on fuel cetane ratings. Many other fuel characteristics, determined by the hydrocarbon mixture, are also important for proper engine operation over a wide range of ambient temperatures. Both exhaust and evaporative emissions are affected by fuel composition, and, in some geographical locations, special "reformulated" fuels have been mandated to reduce emissions. In addition to modifications of basic fuel compositions, additives have been introduced to ensure satisfactory performance, for instance, to improve storage characteristics or reduce the formation of engine deposits.

For the PNGV program, the greatest uncertainty about the CIDI engine meeting its performance targets is associated with emissions, both NO_x and particulates. One way to reduce particulates in the exhaust is to lower the sulfur content of the fuel. As particulate emissions standards become more stringent, lower fuel sulfur levels will almost certainly be required, as well as other measures to control particulate emissions. Lower sulfur content will also be necessary if a petroleum fuel is used as the feedstock for an onboard reformer for a fuel cell, and lower sulfur levels will be beneficial for NO_x treatment technologies. Reducing sulfur content increases the complexity and cost of refining processing. The magnitude of the increase will depend, among other things, on the sulfur content of the crude oil being refined. Nevertheless, a fuel with low sulfur content could be introduced in much less time than a completely new fuel.

Another way to produce low-sulfur fuel is to start with natural gas and synthesize a liquid fuel through the Fischer-Tropsch process. Natural gas is also the starting point for producing DME, a fuel with the potential to reduce both NO_x and particulate levels in CIDI engines. The PNGV program is in the process of testing these and other types of fuel in CIDI engines. Because engine performance may vary with different fuels, joint development and evaluation by the transportation fuel and automobile industries will be important. The potential and time frame for making these fuels widely available must also be assessed.

Fuel cells require hydrogen with low (ppm) concentration levels of CO and sulfur compounds to function efficiently. To avoid energy losses associated with an onboard reformer, the widespread availability of hydrogen will be a critical factor in the practicality of using fuel cells as energy converters. One possible option would be to produce hydrogen in large-scale plants at central locations and distribute it in pipelines. Another would be to convert hydrocarbon fuels into hydrogen at local fueling stations. Substantial energy losses and emissions can occur at various points in the process of making hydrogen available for automotive fuel cells. Sensible trade-offs for this fuel and power plant combination will require that the entire system, from the wellhead to the vehicle wheels, be analyzed. However, trade-offs would require specific criteria, such as reduced petroleum consumption or total energy expended, in addition to the vehicle energy consumption goal of PNGV.

FUEL INDUSTRY

Petroleum is an extremely versatile raw material, and petroleum refinery processes produce a wide variety of products used in many different industries. Refineries vary the proportion and composition of each product to suit market needs, and the economic success of each business entity in the industry depends largely on producing an optimum output mix of products from each refinery. Changing the fuel requirements for new cars will present a major problem for refineries because gasoline generally represents a high proportion of their output. In addition, petroleum refining is a highly capital intensive business, and it generally takes a long time to implement a major change. The fuel distribution system is also very extensive and complex, and changes are difficult to make and require a long time to implement.

If a new automobile requires a unique fuel, both the manufacturer and the transportation fuels industry will be faced with a difficult dilemma. The mobility of the car will be severely restricted if the new fuel is not distributed widely. It must be available even in remote locations. If the availability of the fuel is limited, the car will have limited sales to the general public. At the same time, it would not be economical for fuel suppliers to provide ubiquitous distribution for the initially small number of new cars that will require the new fuel. This problem has limited the success of the government-encouraged "alternative fuel" (methanol, ethanol, natural gas) vehicle programs, and it will continue to prevent the introduction of cars that require radically different fuels unless it is recognized and solved in some creative way. Even the widespread distribution of a new diesel fuel suited to automobiles would be a significant problem in the United States.

Another important characteristic of the transportation fuels industry is that it consists of a large number of companies that vary widely in size, location of facilities, product mix, type of crude oil available, and many other significant factors. Also, it is very common for the product streams from different refineries and even different companies to become mixed at some point in the distribution system. Any significant change in automotive fuel will require a common product specification for all companies. However, this requirement may have widely different economic effects and may make different technical demands on each company. For some, the change may be accommodated easily, but for others the change may spell disaster. Differences in the composition of the crude oil available to different companies could magnify the difficulties of making the change.

The intimate connection between vehicles and fuels has also drawn attention abroad. In Europe, energy companies, vehicle manufacturers, and legislatures have formed a European Auto Oil Program to determine vehicle emission controls for 2000 and 2005, as well as new legislation on fuel quality (Jones, 1997). This program is based on three core principles, namely, linking new legislation to air quality targets, looking at the whole vehicle-fuel system, and choosing the most cost-effective options.

In September 1996, Japan launched the Japan Clean Air Program in coopera-tion with the Ministry of International Trade and Industry, the Petroleum Asso-ciation of Japan, and the Japan Automobile Manufacturers' Association. Program objectives include:

- clarification of mid- to long-term strategies for automotive and fuels tech-nologies that will reduce environmental loads to as low as can reasonably be achieved
- research into the effects of automobile and fuel technologies on vehicle emissions
- the development of the next generation of clean automobile and fuel tech-nologies
- the development of cost-effective automotive and fuel technology mea-sures to improve air quality

PNGV should evaluate the European and Japanese programs that involve the transportation fuels industry when designing an appropriate integrated approach for the United States.

CONCLUSIONS

A widespread supply of a properly tailored fuel is critical to the success of any automotive power plant. A new power plant typically will operate optimally only if it is supplied with fuel tailored to its needs. If future automobiles are expected to require a fuel that is significantly different from the fuels now in use, extensive investigations of the feasibility, economics, and environmental impact associated with its production and distribution should be undertaken early in the development process. A major change in the fuel system infrastructure will re-quire an even longer lead time. If a sequence of changes is anticipated by the PNGV program, the economics of the industries involved should be studied care-fully. A study may indicate, for example, that only one major change will be economically feasible for the foreseeable future. If so, the choice of automobile power plants will be restricted to a sequence that is compatible with a common fuel. PNGV must recognize the tight linkage between the fuel producers and the automobile industry in order for the contemplated changes to be implemented successfully. To increase the PNGV's likelihood of success, a partnership (simi-lar to the PNGV) between the U.S. government and the key automotive fuel pro-ducers should be considered.

RECOMMENDATIONS

Recommendation. The PNGV should propose ways to involve the transpor-tation fuels industry in a partnership with the government to help achieve PNGV goals.

Recommendation. PNGV's choices of energy conversion technologies should take full account of the implications for fuel development, supply, and distribution (infrastructure), as well as the economics and timing required to ensure the widespread availability of the fuel.

Recommendation. The overall societal goals of the PNGV program should be clarified to include potential secondary energy and emissions effects.

6

Major Crosscutting Issues

Eight primary crosscutting issues of the PNGV program are considered in this chapter: (1) the technologies selected for the concept vehicles, specifically the economic viability of the HEV (hybrid electric vehicle); (2) the balance and adequacy of the PNGV program to meet program goals and schedules; (3) major achievements and technical barriers; (4) vehicle safety; (5) developments in foreign technology; (6) goals 1 and 2; (7) government involvement in the PNGV after the technology selection process has been completed; and (8) interactions between the PNGV program and other federally funded research programs.

After the technology selection of specific configurations for concept demonstration vehicles, the development of technologies for low fuel consumption vehicles will take two distinct paths. The first is the concept demonstration, which involves the systematic refinement of systems and components for lightweight, low-loss, CIDI-powered HEVs towards viable production vehicles. The second is the continued development of promising technologies that were not initially selected for the year 2000 concept demonstration vehicles. These technologies will require research studies and feasibility demonstrations until the risks and rewards of their automotive applications are better understood. The concept vehicle demonstration programs will be managed as separate projects by the USCAR partners, whereas the longer-term development will involve many government agencies, suppliers, universities, and government laboratories, as well as the USCAR partners.

TECHNOLOGY SELECTION: ECONOMIC VIABILITY
OF THE HYBRID ELECTRIC VEHICLE

In the past, the committee has been critical of the PNGV systems analysis team for not providing systematic evaluations of candidate technologies in terms of their potential initial costs, the total cost of vehicle ownership, weight impacts, and effects on the infrastructure. The lack of analyses seems to have had a minimal impact, however, on the selection of a configuration for the concept demonstration vehicles because only a limited number of technologies had reached a level of maturity that would justify their selection. It is not surprising, therefore, that all three USCAR partners chose configurations with major weight reductions, reduced aerodynamic drag, reduced accessory loads, low rolling resistance tires, low-loss transmissions, CIDI engines, and parallel hybrid configurations that allow some regeneration of braking energy (see Appendix F). But studies evaluating economic viability for the consumer market have not yet been done.

Vehicles that meet all PNGV targets for aerodynamics, weight reduction, and low rolling resistance are currently projected to achieve less than 65 mpg with any of the nonhybrid combustion engine power trains being considered. Therefore, they would save, at most, about 3,300 gallons of fuel over the lifetime of the vehicle (see Table 6-1). Although 65 mpg falls short of the 80 mpg target, which would nominally save about 3,725 gallons of fuel, the nonhybrid vehicle has the potential to provide 85 percent of the target fuel savings. Assuming that the vehicle meets the PNGV requirements for emissions, safety, size, comfort, range, and driveability, it is appropriate to consider whether it meets the requirement for "equivalent cost of ownership." The principal elements of the cost of ownership are initial cost and lifetime fuel and maintenance costs. This nonhybrid vehicle would have similar complexity to existing diesel power train vehicles, suggesting that they would require no substantial change in maintenance costs and that the initial cost, assuming PNGV component cost targets were met, would increase by less than the value of lifetime fuel savings. Consequently, the total cost of ownership for the nonhybrid vehicle could be lower than the cost for today's baseline conventional vehicle.

Even in the most optimistic projection, a hybrid vehicle is estimated to offer about a 20 percent improvement in fuel economy. For an HEV, the 80 mpg target, therefore, would save less than 425 gallons of fuel over the lifetime of the vehicle over a nonhybrid vehicle with a fuel economy of 65 mpg. But the HEV would be more complex because of its motor/generator, battery, power conversion electronics, and switch gear and electronic controls, and even the most optimistic cost targets for HEVs are still in excess of the fuel savings by approximately an order of magnitude, not including the anticipated additional maintenance costs for the additional components. The cost of ownership would appear to increase markedly for the hybrid vehicle unless these are offset by

TABLE 6-1   Fuel Comparison for the Technologies Selected for the Concept Vehicles

	Baseline	Nonhybrid[a]	Hybrid	Goal
Fuel economy (gasoline equivalent) (mpg)	26.7	≤ 65	≤ 80	80
Fuel used over vehicle lifetime (150,000 miles) (gal)	5,600	≥ 2,300	≥ 1,875	1,875
Lifetime fuel savings over baseline (gal)		≤ 3,300	≤ 3,725	3,725

[a]Vehicle with 40 percent weight reduction, advanced CIDI engine, low drag, low rolling resistance tires, low accessory loads, and low-loss transmission.

overriding benefits related to emissions or operating flexibility. Thus, the hybrid features may not meet the PNGV criteria for equivalent cost of ownership and may ultimately have questionable marketability for this class of passenger vehicles.

A recent review by the NRC Committee on Advanced Automotive Technologies Plan of the U.S. Department of Energy's Office of Advanced Automotive Technologies (OAAT) Research and Development Plan also noted the likely increase in costs and complexity of hybrid vehicles compared to nonhybrid vehicles (NRC, 1998). In that report, the committee also recommended that the OAAT conduct a thorough analysis of the trade-offs between hybrid and nonhybrid vehicle configurations.

Recommendation. The PNGV should continue to refine its detailed analyses of the cost of ownership of hybrid and nonhybrid vehicles. If the economic and performance benefits of the hybrid electric vehicle do not exceed or warrant its additional costs, the concept demonstration vehicle program should be expanded to include nonhybrid vehicles to accelerate the development and introduction of economically viable technologies.

Recommendation. Conventional, nonhybrid vehicles should not be excluded from future PNGV and U.S. Department of Energy plans.

ADEQUACY AND BALANCE OF THE PNGV PROGRAM

As the committee noted in its third report, assessing the adequacy and balance of the PNGV program is difficult because the committee has not seen a funding plan. Based on extensive discussions during this review and the previous three reviews, and on assessments by the PNGV, the committee continues to believe that additional resources will be needed to meet the challenging objectives of the program. However, the fact that the concrete milestone of technology selection in 1997 for a single concept demonstration vehicle configuration has been made (see Appendix F) leads to a number of conclusions.

PNGV's activities to date have been adequate to support the selection of concept demonstration technologies with a high potential of approaching the 80 mpg goal on the PNGV schedule. Although the PNGV estimates that the concept vehicle will fall 10 to 20 percent short of the 80 mpg goal, the improvement will represent a major achievement for either hybrid or nonhybrid vehicle configurations.

The potential for high fuel economy is the result of a number of technical developments. Although the most advanced diesel technology is largely based in Europe, it is available to the USCAR partners through affiliates and agreements. The state of development of hybrid power regeneration components and batteries has been advanced by the HEV and United States Advanced Battery Consortium programs, which are sponsored by DOE and include substantial cost sharing by USCAR partners and suppliers. These programs have provided enabling technology to support concept selection and risk reduction. Similarly, work conducted under PNGV Goal 1 and industry programs have contributed to the development of aluminum and other lightweight materials for major vehicle weight reductions. Assuming that the PNGV/USCAR partners perform as expected (and assuming successful cost reduction), and assuming that the PNGV/USCAR partners can overcome the formidable challenges of emission goals, the committee believes that adequate resources have been applied to the selected technologies and that the year 2000 concept demonstration vehicles and the 2004 prototype vehicles can be realized.

The development of alternative energy converters and storage devices—notably fuel cells, gas turbines, Stirling cycle engines, flywheels, and ultracapacitors—have not progressed to a level where they could be selected for the year 2000 concept vehicles. Furthermore, it is difficult to determine how much, if any, PNGV has stimulated a greater allocation of resources to the development of these alternative systems and devices for automotive applications, with the exception of a recent acceleration of U.S. investment in fuel cells. Observers who expected that the development of these higher-risk concepts would be accelerated by PNGV may be disappointed. But the committee recognizes that there is no assurance that added resources would have brought the risk or payoff of any of these emerging technologies to the point where they would have been selected as superior technologies at this time. The committee also recognizes that PNGV has not agreed on levels of resources or schedules for the ongoing development of these alternative technologies.

The committee believes that one of the primary benefits of PNGV has been to focus various government and industry groups on achieving a revolutionary decrease in automotive vehicle fuel consumption and recognizing of the potential benefits of specific technologies. Looking ahead, it will be important for PNGV to measure progress toward the production and commercial introduction of the selected lower-risk technologies embodied in the concept vehicles, as well as progress toward overcoming the barriers in the development of post-concept vehicle technologies. Growing concerns worldwide about CO_2 levels, including the possibility of an international agreement that would be followed by the implementation of national strategies, could accelerate the need for lower-risk near-term improvements in fuel consumption in the world's automotive fleets. This scenario would also probably mean a larger investment by the U.S. government in longer-term research, as was recommended by The President's Committee of Advisors on Science and Technology in a recent report (PCAST, 1997).

The nonhybrid vehicle uses relatively low-risk technologies and has the potential to reduce the cost of ownership. Although this vehicle will still involve some risk, as well as further significant development, capital investment, and increased infrastructure costs, accelerated implementation of these technologies leading to an approximately 60+ mpg vehicle, coupled with expanded research in the areas specified for the concept vehicles (see the section on Government Involvement in Post-Concept Vehicles below) to achieve an 80 mpg vehicle, seems both feasible and prudent.

The committee believes that technology developed for the PNGV midsize sedan should also be appropriate for light trucks (pickups, minivans, and sport utility vehicles). Light trucks, which have increased to almost 50 percent of sales in the United States, are heavier and consume more fuel per mile than most automobiles. The U.S. Department of Transportation has also expressed concerns (DOT, 1997b) about safety because of the weight differential between light trucks and passenger cars (see the Vehicle Safety section below).

Given the growing popularity of light trucks, the committee believes that the PNGV should identify strategies that would take into consideration these changes in the market. For example, perhaps emphasis on goals 1 and 2 should be increased with the intention of transferring technical improvements in fuel efficiency to light trucks. Or perhaps PNGV should consider parallel programs for automobiles and light trucks. If present trends continue, making an impact on total U.S. transportation fuel consumption through the PNGV will require that PNGV address the issue of light trucks.

Recommendation. Government and industry policy makers should review the benefits and implications of PNGV pursuing a parallel strategy to achieve a 60+ mpg nonhybrid vehicle at an early date and should establish goals, schedules, and resource requirements for a coordinated development program.

Recommendation. The PNGV should assess the implications of the growing vehicle population of light trucks in the U.S. market in terms of overall fuel economy, emissions, and safety. Wherever possible, the PNGV should develop strategies for transferring PNGV technical advances to light trucks.

MAJOR ACHIEVEMENTS AND TECHNICAL BARRIERS

In a presentation to the committee at its October 1997 meeting, the PNGV listed the following important achievements in 1997, in addition to the technology selection and improvements in systems analysis:

- demonstration of a CO-tolerant fuel-cell stack
- integration of a partial oxidation gasoline processor with fuel-cell stack for gasoline-to-electricity demonstration
- completion of a lightweight CIDI engine architecture study
- completion of a DME design study for CIDI engine alternative fuel
- demonstration of lithium-ion 6-Ah cell abuse tolerance
- identification of a potential lightweight approach for containment of low-energy flywheel
- development of an industry-wide specification for integrated power module
- completion of systems analysis for fuel economy trade-offs for major alternative vehicle configurations
- determination of weight reduction potential for selected body structure
- production of driveable hybrid propulsion demonstration vehicles ("mules") at all three companies
- development of continuous casting technology as a replacement for ingot casting, with promising results for reducing the cost of aluminum sheet

All of the achievements listed above, except for the fuel cell and flywheel, apply to the technologies selected for the concept demonstration vehicles. The committee was also apprised of an impressive achievement by Ford, the P2000 concept vehicle, which includes a low-weight, unitized aluminum body into which Ford intends to install their new CIDI engine, the 1.2-liter DIATA (direct-injection, aluminum, through-bolt assembly) engine. Although the P2000 is a Ford program rather than a shared PNGV project, it is representative of the rapid progress the PNGV/USCAR partners have been making, and can make, in the development of fuel-efficient concept demonstration vehicles.

Limited progress was reported under the auspices of the PNGV on technologies beyond those selected for the concept vehicles, except in the area of fuel cells, where progress on stack-power density (abroad) and on fuel reformers (in the United States) has exceeded expectations. In the committee's third report, the PNGV provided a list of major barriers and program needs, which the committee asked the PNGV to update for this report (see Table 6-2). Although important technical advances have been made in the PNGV program, the remaining barriers make the achievement of a vehicle that approaches the 80 mpg fuel economy level and meets all of the other Goal 3 objectives a daunting challenge. Based on a presentation by the PNGV, the committee identified the following principal barriers or technologies for the concept demonstration vehicles:

TABLE 6-2   PNGV's Assessment of Major Barriers and Program Needs

Issues	Challenges
Technical	• control of particulates and NO_x from 4SDI engines • compact fuel-flexible fuel processor for PEM fuel cells • thermal management of lithium battery systems • lightweight, highly efficient power electronics
Production cost	• low-cost lamination material and processing for electric rotors and stators • low-cost aluminum sheet, magnesium, and carbon fiber • low-cost power electronic components and systems • low-cost high-pressure common rail fuel injector and pump • low-cost fabrication processes for batteries and fuel cells
Funding	• cost-sharing for high-risk programs • mobilization of supplier resources • long lead-time from identification of an R&D need to initiation of a government contract • administrative complexity of government programs • nonstrategic distribution of resources
Schedule	• difficulties of meeting all performance and cost targets by 2004 because of continuing funding limitations • recognition that technology selections occur continually throughout the program as data-driven events
Other	• potential changes in emissions regulations • mobilizing the energy industry to enable joint research to achieve emissions/fuel economy targets

Source: Provided by the PNGV in response to a committee request as an update to Table 6-2 from the committee's third report (NRC, 1997).

- meeting overall cost objectives in all major areas
- meeting the stringent emissions target for CIDI engines with an acceptable fuel infrastructure
- developing efficient systems for recovering braking energy and using it to meet peak power demands
- meeting safety targets and extending the life cycle of batteries in HEVs

The committee identified the following principal barriers for technologies beyond the concept vehicle demonstrations:

- reducing fuel-cell system cost and improving reformer performance (including emissions, efficiency, and start-up cycle)
- improving the structural integrity and thermal efficiency for the gas turbine engine
- solving the safety containment problem and reducing the cost of a flywheel energy recovery system

Recommendation. PNGV and the USCAR partners should continue to make safety a high priority as they move toward realization of the concept vehicles.

VEHICLE SAFETY

New structural materials, power plants, fuels (including hydrogen), energy storage devices, and glazing materials are being considered by the PNGV to improve power train efficiency, energy storage, and weight reduction. For every new technology, new failure modes and safety concerns will have to be considered, including crash performance, flammability, explosion, electrical shock, and toxicity. The committee decided not to review safety issues in depth with the PNGV technical teams until the technology selection milestone has been completed but is satisfied that they are aware of many safety issues and are addressing them on an ongoing basis as part of the overall program. For example, failure modes for all promising technologies are being investigated; the safety concerns associated with handling and storing onboard hydrogen for fuel-cell powered vehicles are being examined; and computer simulations are being used to examine the crash performance of hybrid vehicles.

It has been well documented that when vehicles are downsized to reduce their weight, occupant safety is reduced (DOT, 1997b). The PNGV weight reduction targets are not being met by downsizing the vehicle but by using lightweight materials. Consequently, the crush space for frontal, side, and rear impacts will be comparable to today's vehicles. Nevertheless, there is likely to be some reduction in crash performance because a lighter vehicle undergoes a larger rate of velocity change than a heavier one when hitting moveable objects, such as other vehicles or breakaway light posts. New design concepts for improving crash performance may have to be implemented to offset this disadvantage.

If a significant number of lightweight automobiles with high fuel economy penetrate the market, then the increased weight differential between these vehicles and light trucks (pickups, minivans, and sport utility vehicles) could lead to an increase in injuries and fatalities. If the weight reduction technologies being evaluated by PNGV for midsize cars are simultaneously introduced into the marketplace in light trucks, there could be an overall reduction in injuries and fatalities because the weight differential between those trucks and cars in the existing vehicle fleet would be reduced. Also, major weight reductions in other classes of vehicles would have a very large impact on the total fuel consumption of the nation's personal transportation fleet, which is a fundamental goal of the PNGV program. The committee believes the National Highway Traffic Safety Administration should become involved in crashworthiness studies of lightweight vehicles comparable to PNGV vehicle designs.

Recommendation. The PNGV and USCAR partners should continue to make safety a high priority as they move toward the realization of the concept vehicles.

DEVELOPMENTS IN FOREIGN TECHNOLOGY

The USCAR partners monitor foreign technology independently, and the committee has been unable to determine PNGV's or the USCAR companies' views of the importance of developments in foreign technology and product plans except in very limited areas. The following comments on international developments are based mostly on the personal observations and knowledge of the committee members. In spite of progress in the PNGV program, the committee is concerned that the pace and funding of PNGV advanced developments may not be at a level to stay competitive on an international basis. In early January 1998, news releases from individual USCAR members tacitly recognized this fact by announcing aggressive new programs and substantial investments in new technologies. A combination of PNGV and in-house company developments and an assessment of developments in foreign technology no doubt provided an effective stimulus.

The committee believes that the USCAR partners are most attuned to the near-term demands of the market, whereas foreign manufacturers, particularly in Japan and Germany, are more inclined to make larger investments in longer-term, radically new technologies. Last year, the committee cited large Japanese investments in compression-ignition and spark-ignition engines, European progress on small diesel engines, large investments in fuel cells in both Europe and Japan, and Japanese investments in lithium battery technology. This trend has continued, with significant foreign investments in advanced and alternate power trains. Foreign manufacturers have greater incentives than U.S. manufacturers because fuel costs (principally because of higher fuel taxes) are higher in their home markets, and other governments have shown a greater interest in reducing greenhouse gas emissions. Another incentive is competing with the U.S. PNGV initiative directed to better fuel economy, which appears to have spurred increased foreign investment and the initiation of national programs in new automotive technologies. For all of these reasons, the rate of progress overseas continues to be significant in most low fuel consumption technologies. Notable progress in some areas is described below.

Limited production HEVs have been introduced to the Japanese market, and numerous demonstrations have been made in the United States and throughout the world. The United States appears to be ahead in the development of motors, generators, and power conditioning.

Progress on fuel cells has been impressive worldwide, with large German, Canadian, and Japanese investments, as well as successes in U.S. programs. Daimler-Benz and Toyota appear to be planning product introductions. In the fall of 1997, Ford joined a partnership with Ballard and Daimler-Benz to develop fuel-cell technology. The United States appears to have substantially more interest in gasoline reformers than foreign competitors.

Japan reported a 37 percent efficient ceramic gas turbine engine demonstration. U.S. programs to develop automotive gas turbine engines are being canceled or phased out.

Japan is a leading supplier of advanced batteries. Impressive progress has been made on the development of advanced batteries by U.S. programs.

Audi introduced a lightweight, but expensive, aluminum vehicle structure to the market. Ford and Chrysler have constructed impressive demonstration vehicles that incorporate a substantial amount of aluminum and reinforced polymers.

The large-scale production of high-speed automotive diesels in Europe is ongoing, and numerous product developments have been made in Japan and Europe. U.S. programs are focused on engine designs and single-cylinder tests. Test engines are based on European and Japanese technology.

Significant progress was made in Japan on GDI (gasoline direct-injection) engines, where production-ready engines are claimed to have efficiencies approaching those of CIDI engines, but with potentially less severe emissions problems and fueled by existing, low-sulfur gasoline. GDI engines had discouraging technology reviews by the PNGV (see Appendix B). If reports from Japan are accurate, a near-term GDI technology could become available with lower weight and cost than the CIDI engine selected by the PNGV.

PNGV GOALS 1 AND 2

Goals 1 and 2 are open-ended and do not have quantitative targets and milestones. Because the Goal 3 concept demonstration vehicles are focusing on relatively near-term technologies, the distinctions between Goal 3, Goal 1 (increasing competitiveness in manufacturing), and Goal 2 (implementing commercially viable developments from ongoing research on conventional vehicles) are becoming blurred. The development of cost-effective vehicles that meet the targets of Goal 3 will require improvements in competitive manufacturing processes and should result in technologies that can be implemented in the automotive sector. In addition, in-house company developments may satisfy all three goals simultaneously.

In keeping with PNGV's request to the committee (see Appendix C) to concentrate on Goal 3 technology selection, the PNGV did not present cost improvements reflected by activities for goals 1 and 2. The committee realizes, however, that reducing costs both through manufacturing productivity and new technology will be essential to meeting the challenging cost objectives for Goal 3 vehicle systems. More than 30 projects were under way for goals 1 and 2 as of last year, but updates on their status or on new projects were not provided. This raises a serious question as to whether work in productivity, process improvements, and technology innovations are applicable to achieving the cost reductions required for the selected technologies. The committee was informed of significant cost reductions in the fuel-cell bi-polar plate by changing to a stamping process. If other significant improvements have already been identified, it would be useful if they were shared with the committee to bolster its confidence that the necessary work is under way.

Proprietary cost data of two of the USCAR partners were reviewed by a subgroup of the committee. The committee members involved in these reviews were satisfied that these two partners were making significant and appropriate efforts to satisfy the cost analysis requirement.

Recommendation. The PNGV should make sure that the priorities for activities for goals 1 and 2 are consistent with Goal 3 and the overall requirements of the program.

GOVERNMENT INVOLVEMENT IN POST-CONCEPT VEHICLES

The committee was asked to comment on "the role of the government in the PNGV program after the technology selection process is complete." The development of low fuel consumption technologies will take different paths depending on whether or not they were selected for the concept vehicle demonstration configuration.

The USCAR partners have decided to build separate demonstration vehicles rather than build a single, joint demonstration vehicle under the combined sponsorship of the PNGV. Leadership and management of these demonstrators will also be done individually without significant government participation, particularly in light of the relatively near-term technologies that have been selected. However, government support for the continued development of longer-term technologies would be in the national interest and would be consistent with the PNGV's objectives. PNGV should consider a national initiative to establish U.S. leadership in CIDI engines with low emissions, weight, and fuel consumption, as well as programs to accelerate the availability of economical manufacturing methods for low-weight vehicles. Government programs related to fuel infrastructure and vehicle safety also appear to be appropriate for government investment although programs related to other developments important to improving fuel economy, such as the development of low-loss transmissions and accessories, are less appropriate for government involvement.

This issue was addressed in a recent NRC report that concluded that DOE funding and coordination of precompetitive R&D were important to the overall success of developing high-risk automotive technologies with high potential payoffs (NRC, 1998). However, a government role in technology areas where industry is either already doing proprietary work on its own or is likely to undertake proprietary work in the expectation of a return on its investment, was not recommended.

The government should take the leading role in the development of post-concept demonstration technologies. The government's primary role has traditionally been to support the development of longer-term technologies for which industrial development is unlikely because of the high risks and long development horizons. Development of high-thermal-efficiency, low-emission technologies are especially important because of current concerns about greenhouse gases, as well as the potential benefits to U.S. economic competitiveness and balance of payments. Consequently, the committee supports the recommendations of the President's Committee of Advisors on Science and Technology to increase investment substantially in this area (PCAST, 1997).

Specific areas that the committee believes are appropriate for federal research and development are listed below (see also NRC [1998]):

Technologies Selected for the 2000 Concept Vehicle

- accelerated, short-term research on CIDI engine combustion and emissions
- research on manufacturing methods for lightweight structural materials
- analysis of fuel infrastructure issues
- analysis of vehicle safety issues
- longer-term research on electrochemical storage devices
- longer-term research on power electronics technologies

Technologies Selected for the Post-Concept Vehicle

- system development and demonstration of fuel cells with emphasis on cost, fuel reforming, and start-up and transient strategies
- development of enabling technologies for advanced gas turbine ceramic components and systems analysis to determine if further development is warranted
- development of enabling technologies for the containment of flywheels and systems analysis to determine if further development is warranted
- development of advanced batteries with longer life, improved performance, enhanced safety, and lower cost

Recommendation. The government should significantly expand its support for the development of selected long-term PNGV technologies that have the potential to improve fuel economy, lower emissions, and be commercially viable.

PNGV'S INTERACTIONS WITH OTHER FEDERAL RESEARCH PROGRAMS

The committee was asked to "consider and comment on how the PNGV program should interact with other federal research programs." To fulfill this request, the committee had to consider the nature of PNGV and the nature of federal research programs that are considered to be PNGV-related.

The PNGV is a partnership of seven government agencies[1] and three U.S. domestic car companies. Through technical and management committees, PNGV makes recommendations on programs and expenditures but, with few exceptions, does not directly control the direction or expenditures of individual projects. Projects are budgeted and controlled by individual government agencies and by individual USCAR partners and suppliers or jointly under the auspices of USCAR. Some university and DOE national laboratory projects can also be considered PNGV projects, but PNGV does not have the line management structure or budgetary authority necessary to control projects directly. Thus, the primary functions of PNGV are communication and exchanges of ideas among the partners regarding technologies and projects that relate to automobiles, and coordinating, recommending, and planning future projects.

The committee was given to understand that the government's expenditures on PNGV in fiscal year 1997 approached $300 million (NRC, 1997). The PNGV issued an estimate that federal funding for fiscal year 1998 was about $225 million (PNGV, 1998). The government's analysis has not been publicly released, but the committee understands that only about half of this total is budgeted specifically for work related to automotive technology, most of it in DOE's OAAT.[2] The remainder is spent on work by various agencies in fulfillment of non-automotive missions but is considered as PNGV funding because the work is in fields of technology that have possible automotive applications. There is no uniform criterion for including or excluding projects, and a substantial amount of the total probably has marginal applicability to the automotive field.

At the same time, relevant work is probably omitted. For example, all work by the National Aeronautics and Space Administration is omitted from this listing by congressional direction, and significant U.S. Department of Defense activities in related fields may also have been omitted. It is not clear to the committee to what degree PNGV activities are coordinated with government-funded R&D on engines for larger vehicles, such as light trucks, sport utility vehicles, and heavy trucks. Given the increasing importance of light trucks and sport utility vehicles in the U.S. market, coordination between programs, such as OAAT's activities and R&D on CIDI engines for light trucks and heavy-duty vehicles, is important (NRC, 1998).

---

[1]The seven federal government agencies in PNGV are the U.S. Department of Energy, the U.S. Department of Commerce, the U.S. Department of Defense, the U.S. Department of Transportation, the Environmental Protection Agency, the National Aeronautics and Space Administration, and the National Science Foundation.

[2]The OAAT budget for fiscal year 1997 was about $125 million. PNGV-related activities include vehicle systems ($39.7 million); advanced heat engines ($19.1 million); fuel cells ($21.2 million); high power energy storage ($8 million); power electronics and electric machines ($3 million); advanced automotive materials ($13.9 million); and alternative fuels ($2.8 million) (NRC, 1998).

The committee wishes to stress that bookkeeping is not the issue. The issue is how PNGV can achieve its goals, how the public can reap the maximum benefit from government-funded development, and how PNGV can encourage the support of technologies that will have high payoffs for automotive application. PNGV technical teams and committees should continue to monitor emerging developments, both privately and federally funded, and to conduct systems analysis and design studies to identify developments that could contribute to low emissions, low fuel consumption, and low cost of ownership. The PNGV should have unrestricted access to all relevant federal research programs.

The vehicle, subsystem, and component models being created for PNGV could be very useful in the Intelligent Transportation System Program, and in certain National Highway Traffic Safety Administration programs including the development of a vehicle simulator, the development of a vehicle with variable handling characteristics,[3] and the intelligent vehicle initiative, which is in the planning phase and not yet funded.[4] PNGV should continue to establish and maintain communication with these groups and to make PNGV's computer models and system analysis tools available. In addition, because the U.S. Department of Transportation/National Highway Traffic Safety Administration regulates vehicle safety, the National Highway Traffic Safety Administration could appropriately become involved in crashworthiness studies of lightweight vehicles. (See a recent study on vehicle size and weight as related to safety [DOT, 1997b]).

PNGV government managers should continue to interact with other federal research programs by sharing their conclusions regarding high payoff technologies for long-term application to automobiles and by recommending to the government agencies within the partnership redirections or augmentations of ongoing projects that would maximize synergy with PNGV goals. PNGV managers should also recommend that projects that contribute to meeting PNGV goals be included in their future budget requests. The technologies recommended by the PNGV will probably change over time in response to new systems analyses and progress in research investigations.

Recommendation. The PNGV should expand its liaison role for the exchange of technology information among federal research programs that are relevant to automotive technologies and should accelerate the sharing of results among the participants in the PNGV on long-term, high-payoff technologies applicable to automobiles.

---

[3]For example, a vehicle could incorporate rear wheel steering on demand or yaw control to correct skids.

[4]The goal of the intelligent vehicle initiative is to accelerate the development, introduction, and commercialization of driver assistance products to reduce motor vehicle crashes and incidents (DOT, 1997a).

7

PNGV's Response to the Phase 3 Report

In its previous three reviews, the NRC Standing Committee to Review the Research Program of the PNGV made a number of recommendations, which have been documented in published reports (NRC, 1994; 1996; 1997). Some of the recommendations have been carried over from one review to the next, reflecting the committee's concern about how well the PNGV has responded or the importance the committee places on a particular recommendation vis-à-vis the ability of the PNGV program to meet its goals. In the third report, the committee made both specific recommendations for each of the technologies under development and general program recommendations. In general, the committee believes the PNGV has been responsive to many of the recommendations for the technology development areas and systems analysis.

The committee recommended in the third report that the PNGV immediately assess the possible effect of regulatory changes to reduce the atmospheric level of fine particulate matter on the viability of passenger car CIDI engines and modify its program, if necessary. The reduction in emissions will mean that the role of fuels in achieving low emission levels for Goal 3 vehicles is still very important. The PNGV is now taking steps to address the fuel-engine issue. In addition, the committee recommended that the PNGV continue to study infrastructure issues, especially the implications of using different fuels. The PNGV is also moving ahead on analyzing infrastructure issues.

The committee was seriously concerned in prior reviews about the delays and low level of effort devoted to systems analysis. The committee is pleased with the significant progress that has been made in this area since the third review. Attention must now be directed towards creating better cost and reliability models. Several of the subsystem models must also be improved—for example, the models used for fuel cells are overly simplistic for steady-state simulations and cannot be used to evaluate transient response and behavior. The electrical and electronics power conversion devices team has made considerable progress in interfacing with the systems analysis team, but it must provide some models necessary for analysis, such as models for motor/generators, power electronic converters, and control algorithms for series and parallel hybrid drive configurations.

In response to recommendations on fuel cells in the third report, realistic projections of energy efficiency (as compared to an advanced spark-ignition or CIDI engine utilizing the same amount of petroleum in both hybrid and nonhybrid modes) have still not been made. Attempts to find lower cost alternatives for high catalyst loadings, low-cost bipolar plates, and low-cost membrane and electrode assembly designs appear to have been minimal. The committee is encouraged that PNGV has initiated a rigorous cost analysis that could help identify approaches to reducing costs.

Since the third report was published, the PNGV has decided not to pursue the gas turbine as a power plant for the Goal 3 vehicle. Therefore, the committee's recommendations on R&D on gas turbines were not followed. The committee believes limited work on enabling technologies, especially on the development of ceramic components, should be continued.

In the battery technology area, the committee believes PNGV should make more detailed assessments of the safety of the battery systems being developed. The PNGV has not addressed the issue of battery control requirements, and models to simulate cost were not reported to the committee.

The committee recommended in the first and second reports that program management and technical leadership of both government and industry activities be made more effective. This issue was not revisited in the third report or in this report because the committee's viewpoint has not been embraced by the PNGV or USCAR. However, the committee continues to have this concern.

The committee also noted in past reports the importance of obtaining and reallocating federal and industry funds to activities with promising technological potential in the time horizon and needs of the program. In the third report, the committee made the following recommendations:

Recommendation. The PNGV partners (USCAR and the federal government) should immediately develop a schedule of resource and funding requirements for each major technical task. This schedule should show current resources and funding for each major technical task and current shortfalls. Upon completion of this schedule, the PNGV partners should provide a strategy to obtain the necessary resources and funding.

Recommendation. In the event that the PNGV (industry and government) does not obtain or chooses not to increase the resource levels and thereby accelerate the pace of development, the PNGV should reconsider the viability of current PNGV program objectives with regard to performance, schedule, and cost.

In the current review, the committee did not request, and did not receive information on, resource levels and the overall funding profiles of the program. The PNGV's response was expressed in a letter from Mary Good, U.S. Department of Commerce, to Trevor Jones, committee chairman (see Appendix B)

> Your approach in this and earlier reports regarding funding, budgets, and schedule is of concern. Apparently, the Committee's vision of PNGV is one of an engineering project leading to a specific product. In contrast, both government and industry consider the Partnership a "best efforts" undertaking towards stretch goals. First and foremost this is a research, development and technology transfer program . . . We strongly feel that the Peer Review should be focused on the technical issues in the future rather than the funding issues.

Nevertheless, the committee continues to believe that a complex technology program like the PNGV requires a clear understanding of the technical objectives, a program schedule, and a determination and application of required resources to achieve overall program objectives. These three basic requirements are essential and common to all types of programs, whether "best efforts" or otherwise.

The committee has ascertained that meaningful progress has been made in many of the technical areas, and resources have been re-allocated to technologies that appear to be the most promising for the concept vehicles. As the program moves beyond the technology selection for the first concept vehicles, the USCAR partners are expected to focus more resources on the development of the concept vehicles. The USCAR partners have also been working on experimental vehicles. In this review, the committee has focused mainly on the level of effort directed toward individual technology areas and whether sufficient resources are available to meet program goals.

The committee is still concerned that the PNGV has not expressed concern over the competitive implications of foreign technological advances and activities.

Finally, an important recommendation that the committee has consistently made is that the involvement of other U.S. government agencies in PNGV, such as the U.S. Department of Transportation, the National Aeronautics and Space Administration, the U.S. Department of Defense, and the EPA, be increased. Cooperation with the EPA is extremely important in light of emissions targets in general and the stretch research objective for particulates in particular and will require focusing attention on the development of exhaust emissions control technologies, as well as on fuels that will lead to low emission levels. The committee notes that the PNGV has involved the EPA in work on fuel and after-treatment technologies. The systems models being developed by PNGV could be useful to the Intelligent Transportation System Program and other programs at the National Highway Traffic Safety Administration. The PNGV should communicate with these groups and explore how these programs can be advantageous to each other. The committee also believes that National Highway Traffic Safety Administration should be involved in crashworthiness studies of lightweight vehicles (DOT, 1997b).

References

Ando, H. 1997a. Mitsubishi GDI Engine—Strategies to Meet European Requirements. AVL Conference, Engine and Environment, September, 4-5, Graz, Austria.

Ando, H. 1997b. Combustion Control Technologies for Direct Injection Engines. Presentation to the Standing Committee to Review the Research Program of the Partnership for a New Generation of Vehicles. The Savoy Suites Hotel, Washington, D.C., December 2, 1997.

DOE (U.S. Department of Energy). 1996. Transportation Energy Data Book: Edition 16. Prepared by the Oak Ridge National Laboratory for the U.S. Department of Energy under Contract No. DE-AC05-96OR22464. Washington, D.C.: U.S. Department of Energy.

DOE. 1997. Scenarios of U.S. carbon reductions: potential impacts of energy-efficient and low-carbon technologies to 2010 and beyond. Interlaboratory Working Group on Energy-Efficient and Low-Carbon Technologies. Washington, D.C.: U.S. Department of Energy.

DOT (U.S. Department of Transportation). 1997a. Intelligent Vehicle Initiative (Draft Business Plan), October. Washington, D.C.: U.S. Department of Transportation.

DOT. 1997b. Relationship of Vehicle Weight to Fatality and Injury Risk in Model Year 1985-1993 Passenger Cars and Light Trucks. National Highway Traffic Safety Administration Summary Report, DOT HS 808 569, April. Washington, D.C.: National Highway Traffic Safety Administration.

Ford Motor Company. 1997. Ford, Daimler-Benz and Ballard to Join Forces to Develop Fuel-Cell Technology for Future Vehicles. Joint News Release, December 15, Dearborn, Michigan.

Herzog, P. 1996. Future High Speed Diesel Engines for Passenger Cars. Presentation to the Standing Committee to Review the Research Program of the Partnership for a New Generation of Vehicles. National Academy of Sciences, Washington, D.C., November 12, 1996.

Ichikawa, Y., T. Tatsumi, T. Nakashima, I. Takehara, and H. Kobayashi. 1997. Current Status of CGT 302 (Progress in Final Phase). Presented at the Gas Turbine and Aeroengine Congress and Exhibition, Orlando, Florida, June 2-5, 1997.

Iwamoto, Y., K. Noma, O. Nakayama, T. Yamauchi, and H. Ando. 1997. Development of Gasoline Direct Injection Engine. Society of Automotive Engineers (SAE) Paper 970541. Warrendale, Pa.: Society of Automotive Engineers.

Jewett, D. 1997. Ford unveils light car of the future. Automotive News Nov. 3: 46.

Jones, T. 1997. Transportation Technology Trends. Presented at 2020 Vision: The Energy World in the Next Century, Aspen, Colorado, July 5-9, 1997.

Kenny, T., M. Salman, and R. Swiatek. 1997. PNGV Systems Analysis Team Status. Presentation to the Standing Committee to Review the Research Program of the Partnership for a New Generation of Vehicles. USCAR Headquarters, Southfield, Michigan, October 14, 1997.

Kume, T., T. Iwamoto, K. Iida, M. Murakami, K. Akishino, and H. Ando. 1996. Combustion Control Technologies for Direct Injection SI Engine. Society of Automotive Engineers (SAE) Paper 960600. Warrendale, Pa.: Society of Automotive Engineers.

Malcolm, R. 1997. Electrical and Electronics Technical Team Review. Presentation to the Standing Committee to Review the Research Program of the Partnership for a New Generation of Vehicles. USCAR Headquarters, Southfield, Michigan, October 14, 1997.

Nakazawa, N., M. Sasaki, T. Nishiyama, M. Iwai, and H. Katagiri. 1997. Status of the Automotive Ceramic Gas Turbine Development Program—Seven Years Progress. Presented at the International Gas Turbine and Aeroengine Congress and Exhibition, Orlando, Florida, June 2-5, 1997.

NRC (National Research Council). 1992. Automotive Fuel Economy. Energy Engineering Board. Washington, D.C.: National Academy Press.

NRC. 1994. Review of the Research Program of the Partnership for a New Generation of Vehicles. Board on Energy and Environmental Systems and Transportation Research Board. Washington, D.C.: National Academy Press.

NRC. 1996. Review of the Research Program of the Partnership for a New Generation of Vehicles, Second Report. Board on Energy and Environmental Systems and Transportation Research Board. Washington, D.C.: National Academy Press.

NRC. 1997. Review of the Research Program of the Partnership for a New Generation of Vehicles, Third Report. Board on Energy and Environmental Systems and Transportation Research Board. Washington, D.C.: National Academy Press.

NRC. 1998. Review of the Research and Development Plan for the Office of Advanced Automotive Technologies. Board on Energy and Environmental Systems. Washington, D.C.: National Academy Press.

OTA (Office of Technology Assessment). 1995. Advanced Automotive Technology: Visions of a Super-Efficient Family Car. OTA-ETI-638. Washington, D.C.: U.S. Government Printing Office.

PCAST (President's Committee of Advisors on Science and Technology). 1997. Federal Energy Research and Development for the Challenges of the Twenty-First Century (November 5). Washington, D.C.: Executive Office of the President.

PNGV (Partnership for a New Generation of Vehicles). 1995. Program Plan (draft). Washington, D.C.: U.S. Department of Commerce, PNGV Secretariat.

PNGV. 1997. Technical Roadmap (updated draft, October). Southfield, Mich.: PNGV/ USCAR.

PNGV. 1998. President Clinton Supercharges PNGV Initiative, Proposes $50 Million Boost in R&D on Fuel-Efficient Vehicles. Washington, D.C.: U.S. Department of Commerce.

Sissine, F. 1996. The Partnership for a New Generation of Vehicles (PNGV). Report No. 96-191 SPR. Washington, D.C.: Congressional Research Service.

Stuef, B. 1997. Vehicle Engineering Team Review. Presentation to the Standing Committee to Review the Research Program of the Partnership for a New Generation of Vehicles. USCAR Headquarters, Southfield, Michigan, October 14, 1997.

The White House. 1993. Historic Partnership Forged with Auto Makers Aims for Three-fold Increase in Fuel Efficiency in as Soon as Ten Years. Washington, D.C.: The White House.

York, R. 1997. Technology Selection Status. Presentation to the Standing Committee to Review the Research Program of the Partnership for a New Generation of Vehicles. USCAR Headquarters, Southfield, Michigan, October 14, 1997.

4SDI Technical Team. 1997. 4SDI Technical Team Review. Presentation to the Standing Committee to Review the Research Program of the Partnership for a New Generation of Vehicles. USCAR Headquarters, Southfield, Michigan, October 14, 1997.

APPENDICES

APPENDIX
A

Biographical Sketches of Committee Members

Trevor O. Jones, chairman, (NAE) is vice chairman of the board of Echlin, Incorporated, a supplier of automotive components primarily to the aftermarket; chairman and chief executive officer (CEO) of International Development Corporation, a private management consulting company; and chairman, president, and CEO (retired) of Libbey-Owens-Ford Co., a major manufacturer of glass for automotive and construction applications. Previously, he was an officer of TRW, Incorporated, serving in various capacities in the company's Automotive Worldwide Sector, including vice president of engineering and group vice president, Transportation Electronics Group. Before joining TRW, he was employed by General Motors in many aerospace and automotive executive positions, including director of General Motors Proving Grounds; director of the Delco Electronics Division, Automotive Electronic and Safety Systems; and director of General Motors' Advanced Product Engineering Group. Mr. Jones is a life fellow of the American Institute of Electrical and Electronics Engineers and has been cited for "leadership in the application of electronics to the automobile." He is also a fellow of the American Society of Automotive Engineers, a fellow of the British Institution of Electrical Engineers, a registered professional engineer in Wisconsin, and a chartered engineer in the United Kingdom. He holds many patents and has lectured and written on the subjects of automotive safety and electronics. He is a member of the National Academy of Engineering and a former member of the National Research Council (NRC) Commission on Engineering and Technical Systems. Mr. Jones has served on several other NRC study committees, including the Committee for a Strategic Transportation Research Study on Highway Safety, and has chaired the NAE Steering Committee on the Impact of Products Liability Law on Innovation. He holds an HNC (Higher National Certificate) in electrical engineering from Aston Technical College and an ONC (Ordinary National Certificate) in mechanical engineering from Liverpool Technical College.

Harry E. Cook (NAE) received his Ph.D. in materials science from Northwestern University. He is a recipient of the Robert Lansing Hardy Medal and the Ralph R. Teetor Award. He has also received awards from the American Institute of Mining and Metallurgical Engineers, is a fellow of the Society of Automotive Engineers, and a fellow of the American Society of Metals. His career in the automotive industry began at the Ford Motor Company as a senior research engineer and culminated with his position as the director of automotive research at Chrysler Motors. Currently, Dr. Cook is a professor at the University of Illinois in the Department of Mechanical and Industrial Engineering and director of the Manufacturing Research Center. His research experience includes work on phase transformations, friction and wear, automotive product development, and competitiveness.

R. Gary Diaz is the group vice president and chief technical officer for Navistar, where he directs the product development, engineering, reliability and quality group of the truck business. Mr. Diaz was formerly senior vice president of manufacturing and engineering for Case Corporation, an agricultural and construction equipment company, where he was responsible for general management and leading global product development and production. He previously held a number of positions with General Dynamics Land Systems, including division vice president and general manager, Development and Integration Business Unit; vice president, Research Engineering and Logistics; director, Engineering Programs; and engineering manager, Advanced Ground Vehicle Technology. Mr. Diaz participated extensively in the development of the M1A2 Abrams tank, notably in the development of the technology base for system sensors, electronics, communications, and software. He also supervised product development and engineering for the advanced amphibious assault vehicle and the heavy assault bridge. Mr. Diaz received his B.S. in mechanical engineering and his M.S. in engineering from the University of Florida.

David E. Foster is professor of mechanical engineering and director, Engine Research Center, University of Wisconsin, Madison. The Engine Research Center has won two center of excellence competitions for engine research and has extensive facilities for research on internal combustion engines, mainly diesels. Dr. Foster's interests include thermodynamics, fluid mechanics, internal combustion engines, combustion kinetics, and emissions formation. He is a recipient of the Ralph R. Teetor Award and the Forest R. McFarland Award of the Society of Automotive Engineers. Professor Foster is active in a number of committees of the Society of Automotive Engineers. He has conducted research in a broad array of areas related to internal combustion engines. He has a Ph.D. in mechanical engineering from the Massachusetts Institute of Technology.

David F. Hagen is president of ESD, the Engineering Society of Detroit. He spent 35 years with Ford Motor Company, where his most recent position (prior to retirement) was general manager, alpha simultaneous engineer, Ford Technical Affairs. Under his leadership, Ford's alpha activity, which involves the identification, assessment, and implementation of new product and process technologies, evolved into the company's global resource for leading-edge automotive product, process, and analytic technologies. Mr. Hagen led the introduction of the first domestic industry feedback electronics, central fuel metering, full electronic engine controls, and numerous four-cylinder, V6, and V8 engines. Mr. Hagen received his B.S. and M.S. in mechanical engineering from the University of Michigan. He is currently serving on the Visionary Manufacturing Committee of the National Research Council and the board of the School of Management, University of Michigan-Dearborn, and the engineering advisory boards of both Western Michigan University and the University of Michigan-Dearborn.

Harold Hing Chuen Kung is professor of chemical engineering at Northwestern University. He was director of the Center for Catalysis and Surface Science. His research includes surface chemistry, catalysis, and chemical reaction engineering. His professional experience includes work as a research chemist at E.I. du Pont de Nemours & Co., Inc. He is the recipient of the P.H. Emmett Award from the North American Catalysis Society, the Japanese Society for the Promotion of Science Fellowship, and the John McClanahan Henske Distinguished Lectureship of Yale University. He has a Ph.D. in chemistry from Northwestern University.

Simone Hochgreb is an associate professor in the Department of Mechanical Engineering, Massachusetts Institute of Technology. Her research focuses on fundamental and applied problems in combustion and chemical kinetics, with particular focus on applications to transportation, internal-combustion engines, and pollutant emission formation. She has been awarded the Society of Automotive Engineers' Ralph R. Teetor Award, and the General Electric Career Development Award and is the Bradley Foundation Career Development Chair. She holds a Ph.D. in mechanical and aerospace engineering from Princeton University.

Fritz Kalhammer is part-time coordinator for the Electric Power Research Institute's (EPRI's) Strategic Science and Technology Group. He was co-chair of the California Air Resources Board's Battery Technical Advisory Panel on electric vehicle batteries, and he currently chairs a similar panel to assess the prospects of fuel cells for electric vehicle propulsion. He has been vice president of EPRI's Strategic Research and Development and has established the institute's programs for energy storage, fuel cells, electric vehicles, and energy conservation. Before joining EPRI, he directed electrochemical energy conversion, storage and process research and development at Stanford Research Institute (now SRI International), and he conducted research in solid state physics at Philco

Corporation and in catalysis at Hoechst, in Germany. He has a Ph.D. in physical chemistry from the University of Munich.

John G. Kassakian (NAE) is professor of electrical engineering and director of the Massachusetts Institute of Technology (MIT) Laboratory for Electromagnetic and Electronic Systems. His expertise is in the use of electronics for the control and conversion of electric energy, industrial and utility applications of power electronics, electronic manufacturing technologies, and automotive electrical and electronic systems. Before joining the MIT faculty, he served in the U.S. Navy. He is on the board of directors of a number of companies and has held numerous positions with the Institute of Electrical and Electronics Engineers (IEEE), including founding president of the IEEE Power Electronics Society. He is a member of the National Academy of Engineering, and a fellow of the IEEE and has received the IEEE's William E. Newell Award for Outstanding Achievements in Power Electronics (1987) and the IEEE Centennial Medal (1984). He has an Sc.D. in electrical engineering from MIT.

Craig Marks (NAE) is president of Creative Management Solutions. He is also adjunct professor in both the College of Engineering and the School of Business Administration at the University of Michigan and co-director of the Joel D. Tauber Manufacturing Institute. He is a retired vice president of technology and productivity for AlliedSignal Automotive with responsibility for product development; manufacturing; quality; health, safety, and environment; communications; and business planning. Previously, in TRW's Automotive Worldwide Sector, Dr. Marks was vice president for engineering and technology and later served as the vice president of technology at TRW Safety Restraint Systems. Before joining TRW, he held various positions at General Motors Corporation, including executive director of the engineering staff; assistant director of advanced product engineering; engineer in charge of power development; electric-vehicle program manager; supervisor for long-range engine development; and executive director of the environmental activities staff. He is a member of the NAE and a fellow of the Society of Automotive Engineers. Dr. Marks received his B.S.M.E., M.S.M.E., and Ph.D. in mechanical engineering from the California Institute of Technology.

John Scott Newman is professor of chemical engineering at the University of California, Berkeley. His research experience is in the design and analysis of electrochemical systems, the transport properties of concentrated electrolytic solutions, and various fuel cells and batteries. He has received the Young Author's Prize from the Electrochemical Society, the David C. Grahame Award, the Henry B. Linford Award, and the Olin Palladium Medal. He has a Ph.D. in chemical engineering from the University of California, Berkeley.

Jerome G. Rivard (NAE) is president of Global Technology and Business Development, advising business and universities on global business approaches to automotive electronics. He previously held a number of senior management positions with the Bendix Corporation and Ford Motor Company, including vice president for the Allied Automotive Sector of Bendix Electronics Group; group director of engineering for Bendix Electronic Fuel Injection Division; manager of the Bendix Automotive Advanced Concepts Program; and chief engineer for the Electrical and Electronics Division of Ford. Mr. Rivard built an engineering group with skills in electronics, electromechanical devices, fluid-flow control, combustion and power production, and control systems integration. He applied a systems approach to technical discipline management and adopted financial management systems to plan and control engineering projects effectively for maximum return on investment. Mr. Rivard is a member of the NAE and a fellow of the Institute of Electrical and Electronic Engineering and the Society of Automotive Engineers. He received his B.S.M.E. from the University of Wisconsin.

Vernon P. Roan is director of the Center for Advanced Studies in Engineering and professor of mechanical engineering at the University of Florida, where he has been a faculty member for nearly 30 years. He was previously a senior design engineer with Pratt and Whitney Aircraft. Dr. Roan has more than 25 years of research and development experience. He is currently developing improved modeling and simulation systems for a fuel-cell bus program and works as a consultant to Pratt and Whitney on advanced gas-turbine propulsion systems. His research at the University of Florida has involved both spark-ignition and diesel engines operating with many alternative fuels and advanced concepts for both types of engine. Together with groups of engineering students he designed and built a 20-passenger diesel-electric bus for the Florida Department of Transportation and a hybrid-electric urban car using an internal-combustion engine and lead-acid batteries. He has served as a consultant to the Jet Propulsion Laboratory, monitoring electric and hybrid vehicle programs. Dr. Roan received his B.S. in aeronautical engineering, his M.S. in engineering from the University of Florida, and his Ph.D. in engineering from the University of Illinois. He has organized and chaired two national meetings on advanced vehicle technologies and a national seminar on the development of fuel-cell-powered automobiles and has published numerous technical papers on innovative propulsion systems.

Supramaniam Srinivasan obtained his B.S. in chemistry from the University of Ceylon and his Ph.D. in physical chemistry from the University of Pennsylvania. He is internationally recognized for his contributions in electrochemistry, electrochemical energy conversion and storage (with emphasis on hydrogen energy technologies) and bioelectrochemistry. Dr. Srinivasan established electrochemistry/electrochemical technology laboratories at the State University of New York-Downstate Medical Center, Brookhaven National Laboratory, Los Alamos National

Laboratory, Institute for Hydrogen Systems at the University of Toronto, and Texas A&M University. While at Brookhaven National Laboratory, he played a major role in initiating the Fuel Cells for Transportation Program, sponsored by the U.S. Department of Energy. He is author of more than 200 publications, including a book, chapters in books, and review and journal articles. Dr. Srinivasan has been an invited or keynote speaker at several national and international meetings. In 1996, Dr. Srinivasan received the Energy Technology Division Research Award from the Electrochemical Society. From September 1996 to September 1997, he was a visiting professor in chemistry at the Université de Poitiers, France, where he was engaged in research on direct methanol fuel cells. Currently, he is a visiting scientist at the Center for Environmental Studies at Princeton University.

F. Blake Wallace is retired chairman and chief executive officer, Allison Engine Company. He has been involved in engineering and management of high technology gas turbines with United Technologies (Pratt and Whitney), AlliedSignal (Garrett), General Electric (Aircraft Engine Group), and Allison. From 1983 to 1993 he rebuilt the Allison Division of General Motors and served as vice president of General Motors Corporation. He has a B.S. in mechanical engineering from the California Institute of Technology and an M.S. and Ph.D. in engineering science from Arizona State University.

APPENDIX
B

Letters from PNGV

UNITED STATES DEPARTMENT OF COMMERCE
The Under Secretary for Technology
Washington, D.C. 20230

MAY 1 3 1997

Mr. Trevor O. Jones
Chairman
Standing Committee to Review the Research of the
Partnership for a New Generation of Vehicles
National Research Council
2101 Constitution Avenue, NW
Washington, DC 20413

Dear Mr. Jones:

Your third report on the Partnership for a New Generation of Vehicles has been reviewed by both government and industry representatives. We commend the Peer Review Committee for its thorough and comprehensive evaluation and reporting of our program. Your technical expertise and assessments are helpful as we move forward.

We are presently evaluating the opportunities highlighted by each of your recommendations. We look forward to reviewing with you our progress in the program during the 1997 review session, including actions in response to your recommendations.

There is one overriding issue that warrants comment. Your approach in this and earlier reports regarding funding, budgets, and schedule is of concern. Apparently, the Committee's vision of PNGV is one of an engineering project leading to a specific product. In contrast, both the government and industry consider the Partnership a "best efforts" undertaking towards stretch goals. First and foremost, this is a research, development and technology transfer program. This is not to imply the stated goals are considered negotiable. The government is engaged in supporting long-range, high-risk research; industry evaluates technologies to determine how appropriate they might be for the marketplace.

We are operating in an environment of commitment to change and the advancement of relevant technologies. The recent PNGV announcements by the car companies to develop experimental vehicles is evidence of their commitment to the spirit of the Partnership. As you know, none of these experimental vehicles were previously scheduled on PNGV roadmaps or plans. These are indications of a receptivity to "leap frogging" an advanced or higher level of technology. In so doing, the OEMs are significantly increasing the likelihood that the broader goals of PNGV will be realized.

2

The Partnership is well aware that more funding would increase the likelihood of achieving our goals. It is incontrovertible that we are having difficulties arranging additional government funding, even in high priority areas. However, several options being pursued may satisfy program needs within the level funding environment we are constrained to operate in. We strongly feel that the Peer Review expertise should be focused on the technical issues in the future rather than the funding issues.

Once again, we appreciate the hard work and technical insights, recommendations, and assessments the Peer Review team has given us. We look forward to your continued involvement with the Partnership.

Sincerely,

Mary L. Good

Enclosure

cy: Dr. J. Gibbons, Director   Thomas Gale
       Office of Science & Technology Policy   Executive Vice President
                                                 Chrysler Corporation

       John McTague   Arvin Mueller
       Vice President, Technical Affairs   Vice President and Group Executive
       Ford Motor Company   NAO Vehicle Development and
                                                   Technical Operations
                                                 General Motors Corporation

       Members, Operational Steering Group
       DoC
       DoE
       DoD
       DoT
       EPA
       NSF
       NASA

THE DEPUTY SECRETARY OF TRANSPORTATION
WASHINGTON, D.C. 20590

August 1, 1997

Mr. Trevor Jones
Chairman, National Research Council Advisory Group
Partnership for a New Generation of Vehicles
National Research Council
2101 Constitution Avenue, N.W., Room HA-270
Washington, DC 20418

Dear Mr. Jones:

I have reviewed the third report of the National Research
Council's (NRC) Standing Committee to Review the Research
Program of the Partnership for a New Generation of Vehicles
(PNGV).

While the report was issued in the spring of 1997, it reflects
NRC's review of the program in late summer and early fall of
1996, prior to the Fiscal Year 1997 budget approval by
Congress. Accordingly, the report does not reflect the
Department of Transportation's (DOT) PNGV budget or activities
that have been undertaken since the beginning of Fiscal Year
1997 (October 1, 1996). The Research and Special Programs
Administration and the National Highway Traffic Safety
Administration (NHTSA) are in the process of contracting for
the fourth NRC report on the PNGV program. I would like to
take this opportunity to update you on DOT's PNGV program so
that it can be considered in that report.

As I indicated in my March 25, 1996, letter to you, NHTSA
requested $5 million for PNGV in its Fiscal Year 1997 request
to Congress. The House Appropriations Committee approved
$2.5 million of this request, with the following report
language:

> "*Partnership for a new generation of vehicles (PNGV).* --
> The Committee has provided $2,500,000 for PNGV, which is
> $2,500,000 less than requested. Automobile manufacturers,
> in conjunction with the Departments of Commerce, Defense,
> Energy, and Interior, are developing technologies for a
> new generation of vehicles that may be three times more
> fuel efficient than current vehicles. NHTSA's
> participation in this activity is important to address
> critical safety issues; however, this cannot be done until
> the most promising technologies that will go into the PNGV
> are chosen. However, according to a recent National
> Academy of Sciences study, systems analysis for PNGV has
> been delayed by 12-18 months. The study also concludes
> that PNGV does not currently have the necessary systems
> analysis tools to adequately support technology selection,

which is scheduled for 1997. Because of concerns raised by the National Academy of Sciences, the Committee has not fully funded NHTSA's PNGV request. Instead, the Committee has provided sufficient funds to allow NHTSA to begin acquiring the necessary computer equipment to develop advanced computer models that evaluate the crashworthiness of conceptual designs and their safety compatibility with contemporary vehicles. The Committee deferred funding for infrastructure analysis because the department has not made a convincing case for conducting this work without knowing which technologies will be contained in the prototype vehicle."

I am pleased to report that, with this Fiscal Year 1997 funding, NHTSA has purchased the computer hardware and software necessary to support the massively parallel processing capability for safety analysis of PNGV designs using finite element models. In addition, NHTSA has funded the development of finite element models for each of the three PNGV baseline vehicles and for other vehicles representing the fleet.

For the Fiscal Year 1998 budget, NHTSA has requested $2.5 million to continue to develop the advanced computer models to evaluate the crashworthiness of conceptual designs. NHTSA will also begin preliminary construction of an analytical model of the U.S. traffic environment including vehicles, crash modes, and frequencies.

Chapter 6 of your third PNGV report also addresses the importance of infrastructure analysis in the PNGV program. To assist in the evaluation and selection of new technologies for the future vehicles, DOT agrees that PNGV needs to study the potential economic and infrastructure impacts of next-generation vehicles and fuels. As a result, the Bureau of Transportation Statistics has initiated a study, supported by the Volpe National Transportation System Center, to assess the impacts resulting from changes in: (1) vehicle construction and materials, (2) power trains, and (3) fuels. The study is being closely coordinated with the Department of Energy's Argonne and Oak Ridge National Laboratories as well as with USCAR. Lastly, the Intelligent Transportation System Joint Program Office has begun discussions with automakers concerning their research in intelligent vehicle technologies (e.g., crash warning and avoidance) which will have an impact on future vehicle design, operations, and safety.

I look forward to the inclusion of all of the ongoing DOT safety and infrastructure activities in the fourth NRC review.

Sincerely,

Mortimer L. Downey

T. W. ASMUS
PHONE: 576-8004
FAX: 576-2182

November 16, 1995

TO: Allan Murray
        Ford - (313) 594-7303

SUBJECT: Peer Review Question on gasoline SIE Thermal Efficiency

It is highly unlikely that gasoline SIE (homogeneous charge) thermal efficiencies will ever meet the PNGV goal 3 target of 40% (cycle averaged). Current and past efforts aimed at direct injection (stratified charge) SIE have demonstrated significant increases in thermal efficiency over homogeneous charge counterparts, however, will likely fall short of this target. Severe emissions and durability challenges have hampered implementation of this approach.

The open-chamber Diesel remains as the most likely candidate to meet the aforementioned target (current status has peak efficiencies in the 40% range with cycle averaged efficiencies somewhat lower). Well-known emissions, cost and power density challenges persist.

Tom

TWA/sf

APPENDIX
C

Statement of Task and Related Letters from PNGV

STATEMENT OF TASK

This Phase 4 independent evaluation will be directed towards the following tasks:

(1) In light of major technical accomplishments since the third review and technical barriers that remain to be overcome, and the response by the PNGV to previous committee recommendations, examine the overall balance and adequacy of the PNGV research effort to meet the program goals and requirements, i.e., technical objectives, schedules, and rate of progress necessary to meet these requirements.
(2) Examine the PNGV technology selection process including how the PNGV is making choices, and the role of government in the PNGV program after the technology selection process is completed.
(3) Consider and comment on how the PNGV program should interface, if appropriate, with the other federal research programs.
(4) Prepare a fourth peer review report.

Information will be gathered from PNGV representatives on the program's progress and plans for continued research and development. PNGV's technology selection process will be examined.

UNITED STATES DEPARTMENT OF COMMERCE
Technology Administration
Washington, D.C. 20230

June 20, 1997

Mr. Douglas C. Bauer
National Research Council
2101 Constitution Avenue, N.W.
HA-280
Washington, D.C. 20418

Dear Doug:

This letter is to clarify the Partnership for a New Generation of Vehicle's position on how we would like the Peers to treat the issue of "resources".

We want the focus of the 4th Review to be on the technical issues (Are we looking on the right technologies? Do we have the right technical program goals and plans? etc.) and on our rate of progress on these technology issues. We do not want the Peers to judge the PNGV effort as a conventional project where resources are the prime influence on rate of progress. Therefore, we are not specifically asking the Peers to evaluate "resources" throughout the PNGV.

When the Peers feel the rate of progress is inadequate in a particular technology area, we expect the Peers to investigate why and to recommend changes, as they have previously. It would be appropriate for the Peers to investigate the range of potential causes of inadequate progress, including but not limited to, the type and adequacy of the resources.

I trust this helps clarify our views on "resources" and we look forward to a good discussion of the rate of progress of our various technical programs.

Sincerely,

George C. Joy
Chairman, PNGV Technical Task Force

UNITED STATES DEPARTMENT OF COMMERCE
Technology Administration
Washington, D.C. 20230

December 16, 1997

Dr. James Zuccehetto
National Academy of Sciences
2101 Constitution Avenue, NW
Washington, DC 20418

Dear Jim:

I am responding to the question you and Trevor Jones raised as to whether this year's Peer Review of the PNGV program should be extended to look at Goals 1 and 2. As you know, Goal 1 covers advanced manufacturing technologies and Goal 2 covers the application of fuel economy and emission technologies to the current generation of vehicles.

The Partnership does not feel a specific, detailed review of Goals 1 and 2 is needed this year. We feel it is very important to have a thorough and thoughtful review of the Goal 3 technical programs and the technology selection decisions. Many, if not most, of the Goal 1 and 2 results are embedded in the Goal 3 projects, which we reviewed with the Peers. For these reasons we think the focus of this year's effort should continue to be on Goal 3.

Please let me know if I can be of any further assistant to you on this matter,

Sincerely,

George Joy
Chairman, PNGV Government Technical Task Force

cc: Gary Bachula
       Trevor Jones
       PNGV Directors via USCAR

D

Committee Meetings and Other Activities

1. **Committee Meeting, October 13–16, 1997, Southfield, Michigan**
   The following presentations were made to the committee:

   Fuel Cell Vehicle Development
   Neil Otto, Ballard Power Systems

   Ultralight Advanced Composite Hybrids and PNGV
   Amory Lovins, Director, Rocky Mountain Institute

   Ford P2000 Proprietary Session (for Committee and NRC Staff Only) at
     Ford Location

   Opening Remarks
   Trevor Jones, Committee Chair
   Ken Baker, Vice President, Research and Development, GM, for USCAR
   *George Joy (U.S. Department of Commerce), Chair, PNGV Government
     Technology Team*

   Discussion of the Committee's Phase 3 Report
   *George Joy (U.S. Department of Commerce), Chair, PNGV Government
     Technology Team*
   Peter Rosenfeld (Chrysler)
   Ross Witschonke (Ford)
   Ron York (GM)

Technology Selection Status
Peter Rosenfeld (Chrysler)
Ross Witschonke (Ford)
Ron York (GM)
Linda Lance, Office of the Vice President of the United States
Pandit Patil (U.S. Department of Energy), Vice Chair, PNGV Government
Technology Team

Review of Significant Progress and Major Barriers Compared to PNGV Milestones and Future Plans:

Systems Analysis and Engineering
Mutasim Salman (GM), PNGV-USCAR Systems Analysis Technical Team

Body-in-White/Vehicle Weight Reduction
Andy Sherman (Ford), PNGV-USCAR Materials Technology Team
Bill Stuef (Ford) PNGV-USCAR Vehicle Engineering Technical Team

Four Stroke Direct Injection Engines
Rich Belaire (Ford), PNGV-USCAR Four Stroke Direct Injection
Technical Team

Flywheels
Tom Kiser (Chrysler), PNGV-USCAR Flywheel Technical Team

2. **Committee Subgroup Visit to the Chrysler Liberty and Technical Affairs Facility for briefing on proprietary concept vehicle technology, November 10 and 14, 1997, Auburn Hills, Michigan**
    Discussions and presentations on the Chrysler Prowler, CCV, HEV, fuel cells, and technology integration. Attendees from Chrysler included: C.N. Ashtiani, C.E. Burroni-Bird, R.L. Davis, S. Dinda, R.C. Fielding, G.M. Heidecker, T.S. Moore, P.M. Rosenfeld, L.E. Rybacki, and S.T. Speth.

3. **Committee Meeting, December 3–5, 1997, The Savoy Suites, Washington, D.C.**
    The following presentations were made to the committee:

Role of the Government after the Technology Selection Process
Gary Bachula, Acting Under Secretary for Technology Administration,
U.S. Department of Commerce
George Joy, Chair PNGV Government Technology Team, U.S.
Department of Commerce

Interface of the PNGV with Other Government R&D Programs
George Joy, Chair, PNGV Government Technology Team

Fuel Economy Potential of Direct-Injection Stratified-Charge Engines
Hiromitsu Ando, Mitsubishi Motors

Fuel Cells
Alfred P. Meyer, Manager, Transportation Business, International
  Fuel Cells

Update on Government Infrastructure Study
Chris Sloane, PNGV Technical Manager, General Motors
Peter Rosefeld, PNGV Director-Chrysler
Rose Witschonke, PNGV Director-Ford

4. **Committee Subgroup Visit to the Ford World Headquarters for briefing on proprietary cost estimates, December 18, 1997, Dearborn, Michigan**

5. **Trevor Jones and Craig Marks attended the joint meeting of the PNGV Technical Teams on January 28, 1998, Detroit, Michigan**

E

United States Council for Automotive Research (USCAR) Consortia

The U.S. automotive industry, through USCAR, has implemented collaborative projects that directly or indirectly support PNGV objectives. These USCAR consortia include:

- Low Emissions Technologies R&D Partnership (LEP)
- Automotive Materials Partnership (USAMP)
- Supercomputer Automotive Applications Partnership (SCAAP)
- Natural Gas Vehicle Technology Partnership (NGV)
- Advanced Battery Consortium (USABC)
- Vehicle Recycling Partnership (VRP)
- Auto/Oil Quality Improvement Research Program
- Environmental Research Consortium (ERC)
- Low Emissions Paint Consortium (LEPC)
- Electrical Wiring Component Applications Partnership (EWCAP)

APPENDIX
F

The PNGV Technology Selection Announcement

PNGV NARROWS FOCUS IN NATIONAL EFFORT TO DEVELOP ADVANCED, AFFORDABLE AUTOMOTIVE TECHNOLOGIES

January 1998 - The Partnership for a New Generation of Vehicles (PNGV) has completed its selection of technologies considered to be the most promising for achieving the ambitious goals of the Partnership, and will now focus its research and technology development efforts in four key system areas:

- hybrid-electric vehicle drive,
- direct-injection engines,
- fuel cells, and
- lightweight materials.

When this historic partnership was announced by President Clinton and the CEO's of Chrysler Corp., Ford Motor Co., and General Motors Corp. in September 1993, the participants recognized that the development of a new generation of vehicles with up to three times the fuel efficiency of conventional cars was a challenge requiring a national initiative. To improve the probability of achieving needed technology breakthroughs, a large number of promising technologies were initially identified for simultaneous research and development. A major milestone was to narrow the technology development efforts by the end of 1997 and focus resources on the most promising research and development.

Four years into the 10-year partnership, PNGV researchers report solid progress toward developing enabling technologies for affordable, midsize, family sedans capable of achieving up to 80 mpg with very low emissions. The advanced concepts recently unveiled by Chrysler, Ford and GM at the 1998 North American International Auto Show in Detroit reflect this continued progress toward PNGV goals.

"The remarkable, new, fuel efficient, experimental cars rolled out at the Detroit auto show prove that our partnership with the Big Three auto makers is showing results," said Vice President Al Gore, "and that we can protect our environment and meet challenges such as global warming in a way that creates jobs and strengthens our economy. PNGV's selection of these technologies for focused research brings us one step closer to the next-generation cars that will both meet the needs of American families and help us reduce pollution and protect our environment."

Under PNGV, teams of scientists and engineers from 19 federal government labs have been working with their counterparts at Chrysler, Ford, GM (under their U.S. Council for Automotive Research umbrella organization), automotive suppliers, and universities. The backbone of the Partnership—created to provide significant energy security, environmental and economic benefits to the nation— is a coordinated portfolio of hundreds of research projects underway at government, auto company, supplier and university research facilities.

HYBRID ELECTRIC VEHICLE (HEV) DRIVE

Today, almost every vehicle in the world is powered solely by an internal combustion engine. Hybrid propulsion systems have two power sources on board a vehicle. One (such as a fuel cell, internal combustion engine, or gas turbine) converts fuel into useable energy. The second power source, an electric motor powered by an advanced energy storage device, lowers the demand placed on the first power source.

When the two HEV power sources are arranged in parallel, one or both can be used depending on the situation. The electric motor often can power the HEV alone in city driving or over flat terrain. When the hybrid is accelerating and climbing hills, the two power sources can work together for optimal performance. Another advantage is that the electric motor can operate as a generator to slow or stop the vehicle; this captures energy normally lost during braking and "regenerates" it into electricity for later use.

High-power batteries with either nickel metal hydride or lithium-ion technology are the most promising devices to store this energy for later use in powering the electric motor. Hybrids require advanced high-power batteries that are designed for repetitive discharge and recharge over 10,000 times a year as the HEV accelerates, climbs hills, and slows or stops using the brakes.

DIRECT-INJECTION ENGINES

PNGV researchers believe highly-fuel efficient, direct-injection (DI) engines—where the fuel is injected directly into each engine cylinder—show the greatest promise for near-term hybrids. Because the DI engine works in concert with an HEV's electric motor, the engine can be smaller and turned off automatically when not needed, thus increasing mileage and reducing emissions.

Vehicles with today's internal combustion engines are very clean – emitting an average of 95 percent less hydrocarbons, carbon monoxide and oxides of nitrogen than vehicles of the mid-1960s. Nonetheless, PNGV researchers are aiming for even lower emissions from next generation vehicles. Important progress has been demonstrated, and the challenges that remain are being addressed from a full systems perspective, with additional research and development in advanced fuel injection, electronic controls and sensors that optimize engine efficiency, advanced catalysts, advanced emissions traps, and fuels.

Already-efficient DI engines can get better mileage when combustion is triggered by highly-compressing the air-fuel mixture so it self-ignites (i.e. compression-ignition), instead of using spark plugs (i.e. spark-ignition) at lower compression ratios. These compression-ignition, direct-injection (CIDI) engines become an especially attractive primary power source for HEVs when operated with either reformulated fuels (for example, low sulfur fuel now available in California) that help catalytic converters work better at cleaning up pollutants, or new fuels (for example, dimethyl ether or "Fischer Tropsch" synthetic fuels made from natural gas) that produce almost no particulates.

Because the integration of DI engines and new fuels is an important element of the PNGV research portfolio, the Partnership will, in the coming months, be working to ensure an ongoing productive dialogue with the fuels industry to achieve the most workable and affordable solutions for the next generation transportation systems.

FUEL CELLS

Over the longer term, fuel cells could offer the auto industry near-zero emission vehicles with long range, good performance, and rapid refueling. Fuel cells generate electricity directly from a chemical reaction between hydrogen and oxygen, triggered by a catalyst. The required hydrogen can be either carried on the vehicle as a compressed gas, or extracted ("reformed") from a fuel, such as gasoline, methanol, ethanol or propane, carried on-board the vehicle. The electricity produced is used to power a traction motor that drives the wheels. Current research is focused on improving fuel cell size, lowering costs, and developing efficient, compact, and responsive on-board fuel reformers that would provide the needed hydrogen.

LIGHTWEIGHT MATERIALS

Tomorrow's vehicles will contain a mix of aluminum, steel, plastic, magnesium and composites (typically a strong, lightweight material comprised of fibers in a binding matrix, such as fiberglass). To make these materials affordable and durable, research is intensifying on vehicle manufacturing methods, structural concepts, design analysis tools, sheet-manufacturing processes, improved material strength, and recyclability.

Since 1975, the weight of a typical family sedan has dropped from 4,000 pounds to 3,300 pounds. To achieve the Partnership's up-to-80 mpg goal, researchers are working to reduce overall vehicle weight by yet another 40 percent to 2,000 pounds. To achieve this, researchers must reduce the mass of both the outer body and chassis by half, trim powertrain weight by 10 percent, and reduce the weight of interior components.

PNGV'S AGGRESSIVE LONG-TERM GOAL

The Partnership's long-term goal is the development of technologies for new generation, midsize family sedans that get up to 80 mpg; carry up to six passengers and 200 pounds of luggage; meet or exceed safety and emissions requirements; provide ample acceleration; are at least 80 percent recyclable; and provide range, comfort and utility similar to today's models.

The Partnership expects American consumers will buy these vehicles only if they cost no more to own and operate than today's models. Because U.S. gasoline prices are among the lowest in the world, few consumers are willing to pay more for advanced technologies even if they provide greatly increased fuel economy.

CONTINUING TECHNICAL ADVANCEMENTS

While the new concepts recently unveiled in Detroit are impressive, significant additional technology breakthroughs and advancements will be required to achieve the ambitious PNGV goals. Chrysler, Ford and GM are all working on high-mileage concept vehicles to debut in 2000, to be followed by production prototypes in 2004. The government partners and their laboratories will continue to participate in high risk, cooperative research and development with the auto industry to advance critical enabling technologies for possible use in these vehicles. The research and commercial applications resulting from the ambitious PNGV timeframe are stepping stones to the next technological breakthroughs that could yield even greater benefits for the nation's energy security, environment, and economic well-being.

For more information about the PNGV, call John Sargent, U.S. Department of Commerce, 202-482-6185; or Ron Beeber, USCAR, 248-223-9011. Also, check out two sites on the World Wide Web: www.uscar.org and www.ta.doc.gov/pngv.

Acronyms

AGT	advanced gas turbine
Ah	ampere hour
ANL	Argonne National Laboratory
BIW	body-in-white
CIDI	compression ignition direct injection
CO	carbon monoxide
CO_2	carbon dioxide
CRADA	cooperative research and development agreement
DIATA	direct injection, aluminum, through-bolt assembly
DISI	direct injection spark ignited
DME	dimethyl ether
DOE	U.S. Department of Energy
EE	electrical and electronic power conversion devices team
EPA	Environmental Protection Agency
FMVSS	federal motor vehicle safety standards
FRP	glass fiber-reinforced plastic composite material
GDI	gasoline direct injection
GM	General Motors Corporation
GrFRP	graphite fiber-reinforced plastic composite material

HEV	hybrid electric vehicle
HVAC	heating, ventilation, and air conditioning
kWh	kilowatt hour
LANL	Los Alamos National Laboratory
LEV	low-emission vehicle
LLNL	Lawrence Livermore National Laboratory
NO_x	nitrogen oxides
NRC	National Research Council
NVH	noise, vibration, and harshness
ONR	Office of Naval Research
ORNL	Oak Ridge National Laboratory
PEBB	power electronic building block
PEM	proton exchange membrane
PEMFC	proton exchange membrane fuel cell
PNGV	Partnership for a New Generation of Vehicles
PNL	Pacific Northwest National Laboratory
ppm	parts per million
PrO_x	preferential oxidation
R&D	research and development
SNL	Sandia National Laboratory
SWRI	Southwest Research Institute
ULEV	ultra-low emission vehicle
USABC	United States Advanced Battery Consortium
USCAR	United States Council for Automotive Research
4SDI	four-stroke direct injection